Evolution and Speciation in Protozoa

T.J. Pandian
Valli Nivas, 9 Old Natham Road
Madurai 625014, TN, India

CRC Press
Taylor & Francis Group
Boca Raton London New York

CRC Press is an imprint of the
Taylor & Francis Group, an **informa** business
A SCIENCE PUBLISHERS BOOK

Cover page: Representative protozoan species to depict their polyphyletic origin (drawn from different sources; for more details, see Figs. 1.3, 1.4, 1.5, 1.6). Vertical presentation of life cycle in representative protozoans (from Figs. 7.2A, 7.3A, 7.5C, 7.6C). Representative cysts of some protozoans (see Fig. 1.28). Conjugation in ciliates (see Fig. 1.23).

First edition published 2023
by CRC Press
6000 Broken Sound Parkway NW, Suite 300, Boca Raton, FL 33487-2742

and by CRC Press
4 Park Square, Milton Park, Abingdon, Oxon, OX14 4RN

Library of Congress Cataloging-in-Publication Data (applied for)

ISBN: 978-1-032-34751-6 (hbk)
ISBN: 978-1-032-34756-1 (pbk)
ISBN: 978-1-003-32371-6 (ebk)

DOI: 10.1201/9781003323716

Typeset in Palatino
by Radiant Productions

Preface

To sustain heterotrophy and motility, the very diverse metazoans have opted for multicellularity but the protozoans for unicellularity. It may be fascinating to know how this 'wrong' choice of preforming the multicellular metazoan functions employing subcellular organelles has limited species diversity in Protozoa. This innovating thinking has led me to author this book.

"For a long time, the idea of evolution was there among scientists and even with religions like Hinduism. With keywords 'Variations, Struggle for existence and Survival of the fittest by Natural Selection', Charles Darwin established the theory of evolution and its by-product speciation. Subsequently, a large number of publications by microbiologists, botanists and zoologists have confirmed the correctness of Darwin's evolutionary theory. Presently, there are more concerns for species diversity than for evolution. The year 2010 marked the International Year of 'Species Diversity'. This book identifies some life history features of plants 'from algae to angiosperms' and environmental factors that accelerate species diversity and others that decelerate it. Some of these features and factors are known but are not adequately recognized. That requires quantification of the identified factors and features."

"In Shakespearean language, one may say, 'Oh, variation, thy name is evolution'. Hence, the idea of quantification of the identified factors and features may look odd and not possible at a time, when information on *per se* is not known for many species and when taxonomy of plants itself is in a fluid but dynamic state. However, I was a little emboldened, as taxonomy itself represents quantification of species, genus and so on, despite variation(s) among individuals within a species. The onerous task of quantification required much of computer search and a few compromises on the number of some taxa". In addition to website citations, the search included 252 publications covering > 546 species. "Yet, the quantifications may neither be exhaustive nor precise. But the proportions arrived and inferred generalizations shall remain valid. A separate chapter to highlight new findings is not included, as there are too many (shown in italics) of them. The Holy Bible states: "Let your light so shine that people may see your good work and praise the Lord". Being innovative and informative, I earnestly hope that this book stands up to the Biblical statement."

March 2022 **T.J. Pandian**
Madurai 625 014

Acknowledgements

It is with pleasure that I wish to place on record my grateful appreciation to Drs. P. Murugesan and E. Vivekanandan for partly reviewing the manuscript of this book and offering valuable comments. The manuscript was ably prepared by Mr. T.S. Surya, M.Sc. and I wish to thank him for his competence, patience and cooperation.

I sincerely thank many authors/publishers, whose published figures are simplified/modified/compiled/redrawn for an easier understanding. To reproduce original figures from published domain, I welcome and gratefully appreciate the open access policy of Evolution and Development, PLoS Genetics, Quaternary Research. For advancing our knowledge on this subject by their rich contributions, I thank my fellow scientists, whose publications are cited in this book.

March, 2022
Madurai 625 014

T.J. Pandian

Contents

Preamble

The last century has witnessed two most important discoveries that characteristically unify all eukaryotes. The first one is the Calvin cycle (Calvin, 1964) in plants, in which a simple organic molecule like glucose is synthesized from water and carbon-dioxide, using solar energy, i.e. the simple molecules serve as 'batteries' to store solar energy and sustain metabolism. The second one is the discovery of Kreb's cycle (Krebs, 1940), in which simple molecules are decomposed to generate ATPs to sustain metabolism. The last two centuries were dedicated to experimentally demonstrate the past geological climate, in which simple inorganic molecules combined to form amino acids and others, from which life emerged. Thanks to the towering biologists Charles Darwin (1809–1882) and Gregor Johann Mendel (1822–1884), the hypothesis of evolution and speciation was developed, which explains how the (Darwinian) variations or preferably, new gene combinations and their (Mendelian) inheritance have sustained evolution and led to species diversity. Subsequent experiments on microbes, plants and animals by several authors have all brought adequate evidences in support of the Darwinian hypothesis. And the hypothesis has now become a universally accepted theory of 'Origin of Species'. Presently, there are more concerns for species diversity than for evolution. In recognition of its importance, the year 2010 was marked as the International Year of Species Diversity. This innovative series is devoted to discover the causes for biodiversity in eukaryotes. From fragments of relevant information followed by incisive analysis, it has brought to light several new findings.

Many authors consider that 'asexual' or clonal multiplication is derived from sexual reproduction. This is a wrong notion. The blue green algae namely the cyanobionts appeared around 2.5 billion years ago (BYA). Their reproduction is limited to clonal multiplication alone. To tide over unfavorable conditions, they produce cysts called akinetes. Sex was discovered ~ 2 BYA (Butlin, 2002). Subsequently, it was successfully manifested in microbes, plants and animals at different times during the geological past. During the checkered history of evolution, the supplementary role played by clonal multiplication has been progressively reduced. So much so, it occurs in all protozoans (except perhaps *Entamoeba*, Fig. 1.10B), 24% of plants (see Pandian, 2022) and 2% of metazoans (Pandian, 2021b).

Animals are heterotrophs and motiles. In them, motility facilitates mate searching and outbreeding. The ensuing sexual reproduction involves meiosis (segregation) during gametogenesis and recombination at fertilization. Though costly (Stelzer, 2015), meiosis and recombination generate new gene combinations that have led to rapid evolution and speciation. As a consequence, animals are enriched with > 1.5 million species (see Pandian, 2021b). Conversely, plants are autotrophs and sessile. In most of them, the sessility has imposed clonal multiplication. However, ~ 85% of flowering plants (i.e. 77% of all plants) achieve cross pollination by symbiotically engaging motile animals. A quarter of plant species are unable to generate as much as new gene combinations (see Pandian, 2022), as animals can do. This inability of plants has limited their diversity to < 374,000 species (Fig. 1). Nevertheless, plants are more amenable to variety diversity than animals. For example, 1,664 variety/crop species have been generated by farmers during the last 10,000 years (y) history of agriculture. But it is limited to 277 variety/animal species during the last 25,000 y history of domestication (see Pandian, 2022). Hence, plants have compensated the limited species diversity by opting and amenable to variety diversity.

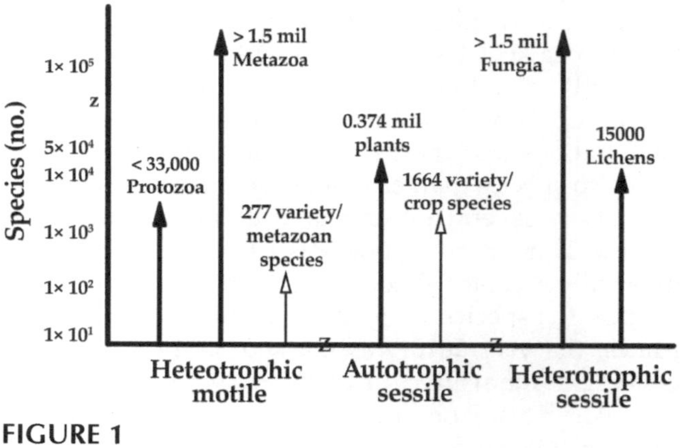

FIGURE 1

Distribution of species number in eukaryotic taxonomic groups mil = million.

In metazoans, the combination of heterotrophy and motility has required the development of different tissues, organs and systems. On the other hand, heterotrophic and motile protozoans have opted perhaps for the 'wrong' choice of performing almost all metazoan functions by specialization and differentiation of subcellular organelles. Briefly, to sustain heterotrophy and motility, metazoans have opted for multicellularity but protozoans for unicellularity. The first consequence of it is the limitation of living species number to 32,950 (see Fig. 1). Another consequence is the secondary loss of sexual reproduction by many protozoans. Approximately, 3,300 species or 10% protozoans can ill-afford sexual reproduction and can multiply clonally alone (Table 5.1). Surprisingly, ~ 1,600 species (e.g. Volvocida, Cryptomonadida, Coccolithsophorida and 50% Dinoflagellida, see Chapter 3) or 5% flagellates are autotrophs but motiles (see Table 3.2). It is from these ancient mastigophores, protozoans and metazoans are considered to have emerged. Barring the unusual proportion of 33.8% parasites (see Table 4.6), the remaining free-living protozoans internally digest captured food.

Surprisingly, the 1.5 million (Hawksworth and Lucking, 2017) speciose Fungia are heterotrophs (Fig. 1), as animals are but sessile, as autotrophic plants are. Unlike animals, they digest the food externally and absorb the digested micronutrients through the body surface. That the sessile Fungi are as speciose as motile animals indicates that species diversity in heterotrophic organisms need not necessarily depend on motility. With adoption of heterotrophism in combination with sessility and multicellularity, the fungi have retained structural simplicity by opting for external digestion and acquisition of micronutrients. Irrespective of sessility, the multicellular 374,000 autotrophic plant species have also developed structurally complex organs (roots, shoots and so on) and systems (vascular system inclusive of xylem and phloem). For example, the number of tissue types increases from 2–3 in algae to 60 in flowering plants but from 6–7 in sponges and cnidarians to > 200 in mammals (Pandian, 2022). Remarkably, the 374,000 speciose sessile plants have also undergone complex structural organization, especially in higher plants clearly indicates that the sessility alone does not deter complex structural organization in multicellular plants. Hence, it is the unicellularity and structural simplicity rather than motility that have limited protozoan diversity to < 33,000 species. Similarly, the adoption of heterotrophism demands complex structural organization, when the organisms choose to engulf larger food and digest it internally, as in animals. However, the other alternative chosen by fungi is to retain structural simplicity and acquire micronutrients after digesting the food externally. In other words, heterotrophism may demand motility and consequent complex structural organization, when the choice of the organism is internal digestion, as in animals or opt for external digestion that may not demand motility and associated complex structural organization, as in fungi.

Lichens are peculiar organisms and represent the symbiotic association between the sessile heterotrophic fungi and autotrophic algae (Fig. 1). Their diversity is limited to ~ 15,000 species (*Wikipedia*). In a way, they are the terrestrial equivalents of the aquatic combination of sessile heterotrophic cnidarians and autotrophic zooxanthellae. The coral – zooxanthellae are widely distributed in tropical and subtropical seas around the earth up to the depth of 200 m. But that of lichens is limited between certain levels of altitude in montane ecosystem (e.g. 3,100 and 3,400 m in Nepal, Baniya et al., 2010, Fig. 2B), albeit they are ubiquitous. In the context of trophic dynamics, the autotrophic plants are producers; they occur and thrive in marine, freshwater and terrestrial habitats. So are metazoans, which serve as consumers. The distribution of protozoans is, however, limited mostly to marine and freshwater habitats (Fig. 2A). Being decomposers, fungi are more common on land than in water. The reverse is true for bacteria (Fig. 2B). It remains to be known whether the Fungi are indeed less common in aquatic systems, as we know less of them from aquatic ecosystems. On the whole, the objective theme of this innovative book is to explore the causes for the limited species diversity in protozoans.

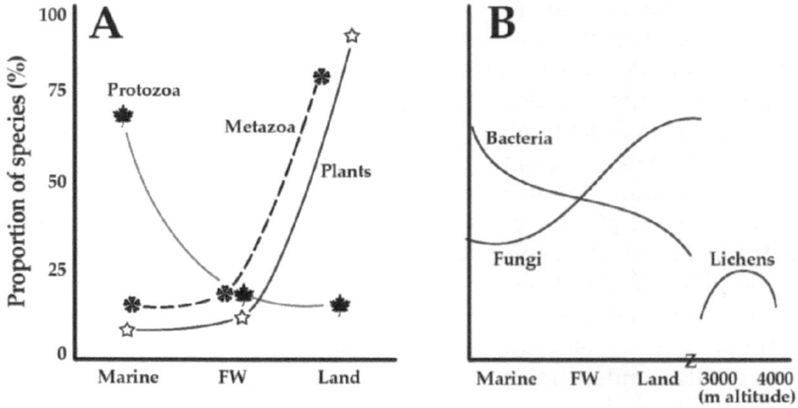

FIGURE 2

(A) Proportion of species distribution in marine, freshwater (FW) and terrestrial habitats. (B) Proportion of bacterial and fungal distribution in marine, freshwater and terrestrial habitats. The right end shows the distribution of lichens at different altitudes in montane systems.

Summary: 1. In heterotrophic motile Protozoa, unicellularity has limited diversity to < 33,000 species. 2a. Sessility has imposed clonal multiplication in > 24% autotrophic plants. b. It has reduced diversity to 374,000 species, c. which is partially compensated by their amenability to variety diversity. d. However, multicellularity in them has facilitated structural complexity. 3a. In motile multicellular Metazoa, the combination of heterotrophy and motility has led to diversity of > 1.5 million species. b. Further, the need for food capture and internal digestion have demanded complex structural organization. 4. In contrast, that of heterotrophy and sessility has also facilitated diversity to > 1.5 million species in Fungia, which have opted for external digestion and absorption of micronutrients; in its turn, this has led to the retention of relatively simpler structural organization than plants and metazoans.

1

General Introduction

Introduction

Protozoa are microscopic unicellular rather acellular eukaryotic animals with complex internal organelles. Metazoa are made up of cells (e.g. 40 trillion cells in human, *Wikipedia*) that are organized into tissues (6–7 tissue types in sponges and cnidarians and > 200 in Mammalia, see Pandian, 2021b), organs and systems. A tissue consists of many cells that are similar in form and function. An organ is composed of different tissue types and the system, in its turn, bear a few organs. It is interesting to know how the few subcellular organelles of protozoa perform almost all functions of metazoa. Protozoa are a large collection of animals with considerable morphological (e.g. Warren et al., 2016), physiological (Horan, 2003) and genetic (Lynn, 2010b) diversity and are assembled together in a phylum for taxonomic convenience. They occur and thrive in diverse habitats of all aquatic systems and also abound in damp soil. They can be photoautotrophics (phytomastigophoran flagellates), phagotrophics, suspension feeders (Nielsen et al., 2017, Thurman et al., 2010) or osmotrophics in all parasites and cellulase secreting symbionts (e.g. Fuente et al., 2006).

Some protozoans, that are of academic and economic importance, are listed hereunder: 1. With the fastest rate of reproductive multiplication, their doubling time ranges between once every 6 hours (h) in ciliates *Cyclidium* and 60 hours in *Keronopsis* (Wang et al., 2005). 2. As a consequence, their abundance in freshwater plankton can reach densities of 10^5–10^6 cell/ml (Sanders, 2009). Oceans provide habitats for ~ 72% of all protozoa (see Finlay and Esteban, 2018). In the open ocean, maximum densities of planktonic radiolarians alone reach up to 10,000/m^3 in the subtropical Pacific (Anderson, 2001). Some benthic radiolarians can be found down to the depth of ~ 8,000 m, although most of them are abundant and diverse between 200 and 2,000 m (Anderson, 1983). The protozoan density ranges from 3 x 10^4 cell/l to 19.65 x 10^4 cell/l in the Xixi wetland, Hangzhou, China, of which 69% are ciliates (Shi et al., 2015). Also on land, their density can be 2.4 x 10^3 in the rhizosphere of one gram soil (see Pandian, 2022). In one gram of ocean sand, 50,000 sedimented foraminiferan shells can be found (see Hyman, 1940). Symbiotic ciliates like *Dasytricha*, *Endodinium* and *Polyplastron* also abound in the rumen liquor mostly at the density of 10^5 cell/ml. When expressed in global terms, their numbers are very large (Finlay and Esteban, 2018) and their immense role in ecosystem functioning is inevitable.

3. Protozoa are small in size (< 20 µm in flagellates, < 50 µm in amoeba, < 200 µm in ciliates, Finlay and Esteban, 2018). However, by sheer abundance, they play a vitally important role in the food web by transferring microscopic organisms to the next higher trophic level. (a) In aquatic system, the presence of excessive nutrients causes eutrophication. A survey indicates that 28, 41, 48, 53 and 54% aquatic systems are eutrophic in Africa, South America, North America, Europe and Asia, respectively. Microalgae rapidly absorb the excessive micronutrients at high efficiencies up to 100%. The acquisition process considerably reduces eutrophication and thereby cleanses the eutrophication, as it otherwise costs US$ 2.2 billion in USA alone (see Pandian, 2022). As the turnover rate of microphagous protozoans is 1.9 times faster than that of metazoans, the bacteriophagous protozoans decelerate eutrophication more rapidly than metazoans (Finlay and Esteban, 1998). (b) The cyanobionts are capable of directly fixing atmospheric nitrogen. Hence, they play a key role in enriching the aquatic system with nitrogen supply. Again, the herbivorous protozoans play an important role by transferring nitrogen fixing bacteria/algae in the food web to higher trophic level in natural aquatic systems. (c) In man-made waste water treatment, protozoa also play a vital role. Being suspension feeders, some sludge inhabiting ciliates, for example, filter bacteria and solids of ultra-microscopic sizes between 0.3 and 1.5 µm (Horan, 2003). Consequently, they generate crystal clear water in activated sludge with significantly reduced number of fecal bacteria and others, as listed below:

Parameter	Without ciliates	With ciliates
BOD (mg/l)	53–70	7–24
Organic nitrogen (mg N/l)	14–21	< 10
Suspended solids (mg/l)	86–118	6–34
Bacterial count (10^6/ml)	106–160	1–9

4. Unusually, > 33% protozoan species are parasites (see Chapter 4). Due to detrimental diseases like malaria, trypanosomiasis, some of them cause ill effects on the wellbeing of man, livestock as well as fishes. 5. Being sensitive indicators of ancient environment, foraminifers serve as ecological indicators in petroleum exploration (Mikhalevich, 2021) and rise in sea levels (Gehrels, 2013). With the key role played in natural ecosystem and man-made

wastewater treatment plant, medical biology and geology, protozoans are academically and economically important animals.

Besides these, the curiosity to know how a single celled protozoa performs the functions of metazoa has attracted great attention. Not surprisingly, 64,468 publications have been generated (*researchgate.net*), i.e. considering 80,650 living and extinct protozoan species (see Table 1.15), each protozoan species has received an average of eight publications. Expectedly, a large number of books are available. But most of them are relegated to specific protozoan groups like Foraminifera (Sen Gupta, 2003), Radiolaria (Anderson, 1983) and Ciliophora (Corliss, 1979), while others are limited to medical biology (Wiser, 2010, Pablos Torro and Morales, 2018). Books of Patterson (1998) are concerned with only freshwater Protozoa but others like Capriulo et al. (1990) only with marine protozoa. There is hardly any book on Protozoan Biology (e.g. Kharna, 2004) but that too is limited mostly to the description to life history of selected species. In this book, an attempt is made to collect, collate, sequence and synthesize relevant information on protozoan biology on a holistic approach from the angle of evolution and species diversity.

1.1 Form and Function

Barring lobose rhizopods, all other protozoans exhibit a definite shape and size. Their shape can be spherical (e.g. *Noctiluca*, Fig. 1.3F), stellate (e.g. *Acanthometra*, Fig. 1.4K), oval (e.g. *Giardia enterica*, Fig. 1.3N), oblong (e.g. *Euglena gracilis*, Fig. 1.3H) or dorsoventrally flattened (e.g. ciliate *Urostyla*). Their size ranges from 2–4 µm in flagellates (Finlay and Esteban, 2018) to 4.5 mm in the freshwater ciliate *Spirostomum ambiguum* (Hyman, 1940) and to > 20 mm (shell) in a Foraminifera. The mean size ranges from 2 µm to 2 mm – amounting to a size range of one thousand times; only ~ 8% aquatic metazoans cover a similar range of body length (Fenchel and Finlay, 2006). The pseudopodial network in foraminifers may reach 20 mm and the spines may increase the size to > 20 mm and make the stellate radiolarians larger than some marine metazoans. Colonial radiolarians, which occur as gelatinous cylinders, may reach 3 m in length (Finlay and Esteban, 2018). The lobose rhizopods and foraminifers (as listed in 130 shapes by Mikhalevich, 2021) are asymmetric and anaxial. But all others are bilateral.

The protozoan body is enclosed with an outer clear gelatinous layer of ectoplasm clothed by a pellicle and inner mass of endoplasm. In the ciliate *Blepharisma*, the pellicle removal does not affect its shape or behavior, and is quickly reformed (see Hyman, 1940). In it, the contractile fibrils, the myonemes are also located, especially in ciliates and gregarines. Situated below it, the endoplasm exhibits a great deal of difference. It is a fluid of granular mass composed of hyaloplasm with various organelles and inclusions. From it, the locomotor organelles, the cilia or flagella are projected from the basal bodies. The other organelles include one to many nuclei, mitochondria, microsomes, Golgi bodies and others. The inclusions comprise food and contractile vacuoles.

Food vacuole: Some protozoa may have a mouth or cytostome (Fig. 1.1A), leading to the gullet (Fig. 1.1B) and tubular or funnel-shaped cytopharynx (Fig. 1.1C). These organelles are developed by an extension of the ectoplasmic surface deep into endoplasm. Food captured directly or through cytostome and passed on to cytopharynx is held eventually in the food or gastric vacuole. Digestion is initiated at acidic phase down from 6.9 to 4.3 pH and subsequently, at alkaline phase, during which most or all digestion is accomplished. Pepsin-like protein is known from rhizopods. Trypsin-like protein is secreted by many protozoa. In them, amylase is also present. Cellulase is reported from symbiotic ciliates. Injected olive oil or fat droplets are ingested and digested by amoebae (Hyman, 1940). In flagellates and ciliates, the indigestible remnants are ejected through cytopyge (Fig. 1.1C).

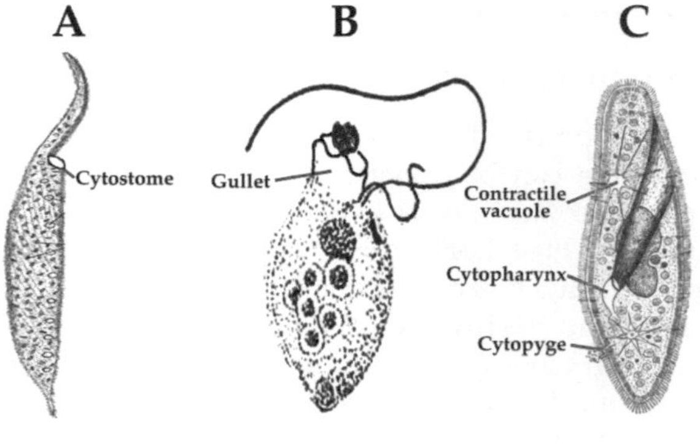

FIGURE 1.1

(A) The ciliate *Dileptus gigas*. (B) The flagellate *Monas vestita* (after Reynolds, 1934). (C) *Paramecium caudatum* (free hand drawings, after Hyman, 1940).

Contractile vacuole is characteristic of freshwater protozoans. It is rhythmically filled and discharged of its contents to the exterior directly or via the cytopyge. The interval between successive discharges may last from a few seconds (s) to several minutes (min). The duration required to discharge fluids equivalent to body volume runs from ~ 9 hours in *Amoeba proteus* to ~ 32 hours in *Paramecium caudatum* (Laybourn-Parry, 1984). The fluid withdrawn from the vacuole of *Spirostomum* is reported to contain urea. Hence, the vacuole, besides being osmoregulatory, may also represent 'the primitive excretory system' (Hyman, 1940).

The protozoan nucleus is vesicular and massive. Its shape can be spherical, oval or biconvex. The Superclass Ciliophora is characteristically binuclear, the generative micronucleus and vegetative macronucleus. The latter usually degenerates during conjugation and is regenerated in the ex-conjugants (see Fig. 1.23). The number of macronuclei can be one to many; for example, it is 80 in *Stentor coeruleus* (see Hyman, 1940). Some photoautotrophic flagellates are endowed with one broad cup-shaped (e.g. *Haematococcus*) or usually a pair of oblong chloroplasts. Their color may be green or golden brown in Dinoflagellida and some Chrysamonadida but olive green, brown, red or blue in some Cryptomonadida (Levine et al., 1980).

With mitochondria, most Protozoa are aerobics. However, many of them are microaerophilics and look for habitats with a low level of dissolved oxygen, which is just adequate to derive the aerobic respiration and low enough to exclude metazoan competitors and predators. They occur and thrive in microaerial habitats like sediments with rich microbial populations. For example, the ciliate *Loxodes* prefer to migrate to low oxygen containing sediments. The most common anaerobic heterotrophic free-living flagellates are *Hexamita, Trepomonas, Retortamonas* in anoxic waters and human parasites like *Giardia intestinalis*. Among Amoebidae, *Mastigamoeba* and *Pelomyxa* do not have mitochondria. Some of them have hydrogenosomes (e.g. *Psalteriomonas, Pseudotrichomonas, Ditrichomonas,* see Finlay and Esteban, 1998), the anaerobic derivatives of mitochondria that produce molecular hydrogen and ATP in anaerobes. Ciliates like *Metopus* and *Caenomorpha* are obligate anaerobes found only in anoxic sediments and hypolimnion of eutrophic waters (Sanders, 2009). The anaerobic protozoa are probably the only phagotrophic organisms capable of living permanently in anoxic waters (Fenchel and Finlay, 1995). Briefly, microaerophilism is a pre-adaptive feature (Finlay and Esteban, 2018) that could be a reason for the highest (33%) proportion of parasitic species among protozoans.

Locomotor organelles: On the basis of their locomotory organelles, Protozoa have been classified. They are characterized by finger-like pseudopodia in Rhizopodea, whip-like filamentous flagella in Mastigophora, hair-like cilia in Ciliophora and with none of them in Sporozoa (Horan, 2003). Within Rhizopoda, four types of pseudopodia are recognized (Hyman, 1940): (i) Lobopods are temporary, continuously changing tubular balloon-like extensions of plasma membrane, into which ecto- and endo-plasms stream (e.g. *Amoeba dubia*, Fig. 1.4A). (ii) Filopods are slender with pointed tips and is composed of ectoplasm alone (e.g. Filosia: *Gromia*, Fig. 1.4F). (iii) Reticulated rhizopods are thread-like pseudopodia that branch and anastomose into a network (e.g. Foraminifera: *Rotalia*, Fig. 1.4G) and (iv) Ray-like axopods are stiffened by central axial rods (e.g. *Acanthometra*, Fig. 1.4K). Using lobopods, amoeba can only creep or crawl over solid substratum. Sporozoa were recently reported to have tiny undulating cell membrane spreading to a forward gliding motion (*parasite.org.au*).

Being the filamentous extension of plasma membrane, flagella undulate to propel through aqueous medium. Their number varies from one (e.g. *Noctiluca,* Fig. 1.3F) to four (e.g. *Trichomonas buccalis,* Fig. 1.3Q) but usually to two. These two may be equal in length and similar in function (*Chilomonas,* Fig. 1.3D). Being the extensions of plasma membrane, cilia are arranged in rows called kinetics. The kinetic arrangement varies from a dense one covering the entire body of the ciliates (e.g. *Opalina,* Fig. 1.6A) to a sparse distribution (e.g. *Vorticella,* Fig. 1.6J). In some ciliates, the tuft of cilia is compactly fused to form compound organelles, such as the cirri that can be used as walking legs (e.g. *Cycloposthium,* Fig. 1.6R) or membranelles that direct water current toward the cytostome (e.g. *Stentor,* Fig. 1.6N, Sanders, 2009). The structural organization is similar in the relatively long flagellum and shorter cilium. But, interconnecting basal elements facilitate a synchronous ciliary beat in the same direction in ciliates (Horan, 2003) to propel the animal to swim through aqueous medium and/or to capture food (*parasite.org*). Both the cilium and flagellum consist of a stiff, elastic, straight or spiral axial fibril, namely the axoneme that is enclosed by the outer plasma membrane with an inner central core of microtubules arranged in '2 + 9' configuration (Fig. 1.2). The configuration consists of two single central microtubules surrounded by nine peripheral doublets. Amazingly, it is conserved across all eukaryotes including the human sperm (*parasite.org.au*). The so called canonical '2 + 9' microtubular axoneme is the principal feature of motile cilia and flagella and is one of the most iconic structures in cell biology (Ginger et al., 2008).

Though both the flagellum and cilium are similar in structure, they differ in the plane of their motility. The motion of flagellum is an undulatory wave, which begins at the base and proceeds through the length of the flagellum and most flagella move only in a planar mode. However, the movement can be helical in some species like *Euglena*. The ciliary motion is characterized by a stroke or beat, which involves the ciliary bending at the base, while the rest of the cilium remains straight. The cilium is subsequently drawn back to its initial position close to the cell surface (Horan, 2003).

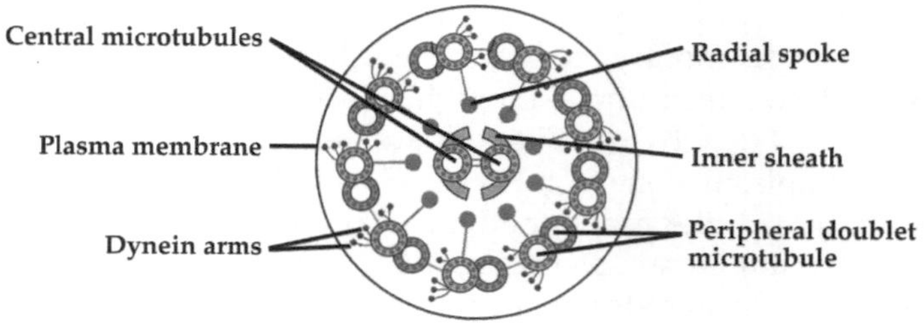

FIGURE 1.2

Ultrastructural organization of a cilium/flagellum (free hand drawing based on Lodish et al., 2000).

1.2 Classification and Taxonomic Distribution

The classification of Protozoa remains in an extremely perplexed problem. Table 1.1 lists some examples for the number of classes and orders erected within the four major groups of Protozoa. Four to 59 orders in Ciliophora are differently named and there is no synonymity of nomenclature even at the class level. Hence, a visitor or an expert may be totally baffled by the confusing classification of Protozoa. Among many, the following two reasons may be traced to the present status. 1. Barring the natural monophyletic Ciliophora (Sanders, 2009), all the remaining protozoans are combined into the three polyphyletic superclasses for taxonomic convenience under the phylum Protozoa. 2. With regard to the subclasses Phytomastigophora, Coccolithophorida, Choanoflagellida, Euglenida and Dinoflagellida are

TABLE 1.1

Number of Classes and Orders listed for the four superclasses in Protozoa

Superclass	Hyman (1940)	Levine et al. (1980)	Microbe-notes. com	Others
Mastigophora	10 Orders in 2 Classes	15 Orders in 2 Classes	17 Orders in 2 Classes	19 Orders in 5 Classes of Ciliophora
Sarcodina	5 Orders in 1 Class	38 Orders in 5 Classes	3 Orders + 6 Subclasses in 3 Classes	59 orders in 11 Classes and the Phylum Ciliophora (Lynn, 2010a)
Sporozoa	8 Orders in 3 Subclasses	36 Orders in 7 Classes	1 Order + 4 Classes	
Ciliophora	4 Orders in 4 Classes	18 Orders in 3 Classes	16 Orders in 5 Subclasses	

considered as algae by Guiry (2012) but protozoans by Hyman (1940) and Levine et al. (1980). As a result, some like Lynn (2010a) have assemblied Ciliophora as a phylum but the others treat them as a class or superclass within the Phylum Protozoa. Nevertheless, this account classifies them into four major superclasses (Tables 1.2 to 1.4).

Mastigophora are Protozoa that are flagellated in most of them but a few may bear the flagella only temporarily. They are classified into 2 classes; the autotrophic/mixotrophic Phytomastigophora consist of 8 orders and heterotrophic, symbiotic or parasitic Class 2 Zoomastigophora (Table 1.2). More number of orders are built, as and when the life history is discovered for more and more species, following the development of cultivation techniques (e.g. Eikrem et al., 2017). Of 6,900 mastigophores, ~ 5,032 species or 73% are free-living flagellates. The following mastigophores are biologically interesting and are mentioned hereunder: (i) unusual of

TABLE 1.2

Superclass (flagellates) 6,900 speciose Mastigophora (compiled from Levine et al., 1980[†], Corliss, 2001, Warren et al., 2016*, *parasite.org*). A = Autotroph, M = Mixotroph, H = Heterotroph, FL = Free-living

Taxon	Name	Characteristics	Species (no.)
Class 1	Phytomastigophora	Photoautotrophs	~ 4845
Order 1	Volvocida	Colonials – 2 flagella, A[†]	400[h]
Order 2	Chloromonadida	Small – amoeboid – 2 flagella, A[†]	500[j]
Order 3	Coccolithophorida	A[†]	46/90[a]
Order 4	Cryptomonadida	Amoeboid – flagella, A[†], H*	> 100[b]
Order 5	Dinoflagellida	2 flagella, 1 massive nucleus, H, M*	~ 2000[c]
Order 6	Chrysomonadida	Amoeboid – flagella – cyst formation, H, M*	?
Order 7	Euglenida	Largest – 2 flagella – cytostome, H*	> 1000[e]
Order 8	Heterochlorida	H*	?
Class 2	Zoomastigophora	Heterotrophic – mostly parasitic	~ 2055
Order 9	Rhizomastigida	Permanently amoeboid – long flagellum, FL	?
Order 10	Choanoflagellida	Collar encircling a flagellum, FL	125[e]
Order 11	Kinetoplastida	Elongated body – anterior flagellum	> 30[f]
Order 12	Bisosoecida		?
Suborder 1	Bodonina	Free-living	32[i]
Suborder 2	Trypanosomatina	Vascular parasites	< 100
Order 13	Hypermastigida	Numerous flagella – gut parasites	
Order 14	Diplomonadida	2 nuclei – 2 flagellar – gut parasites	
Order 15	Retortamonadida	Oblong – 2 flagella – cytopharynx	1800[g]
Order 16	Oxymonadida	Oblong – 4 flagella	
Order 17	Trichomonadida	Spherical – 2 flagella	

a = /Okada and Honjo (1973), b, d, j = *onezoom.org*, c = /Taylor et al. (2008), e = *wikipedia*, f = d'Avila-Levy et al. (2015), g = Adlard and O'Donoglue, 1998, h = see Section 1.4, i = Guiry, 2012

flagellates, *Rhynchomonas nasuta* is found in soil as well as in freshwater and marine habitats (Lee and Patterson, 1998). (ii) *Hexamita* is anaerobic, and inhabits anoxic sediments. (iii) *Noctiluca* (Fig. 1.3F) is noted for its phosphorescence and is responsible for most of the phosphorescence in coastal waters. (iv) *Haematococcus pluvialis*, appear suddenly in pools, produce the mythical red rain (Hyman, 1940). (v) The 4-flagellated *Janickiella* and *Trichomitus* (Fenchel and Finlay, 1995) as well as the 6-flagellated *Streblomastix* degrade cellulase in the hindgut of termites and wood-eating

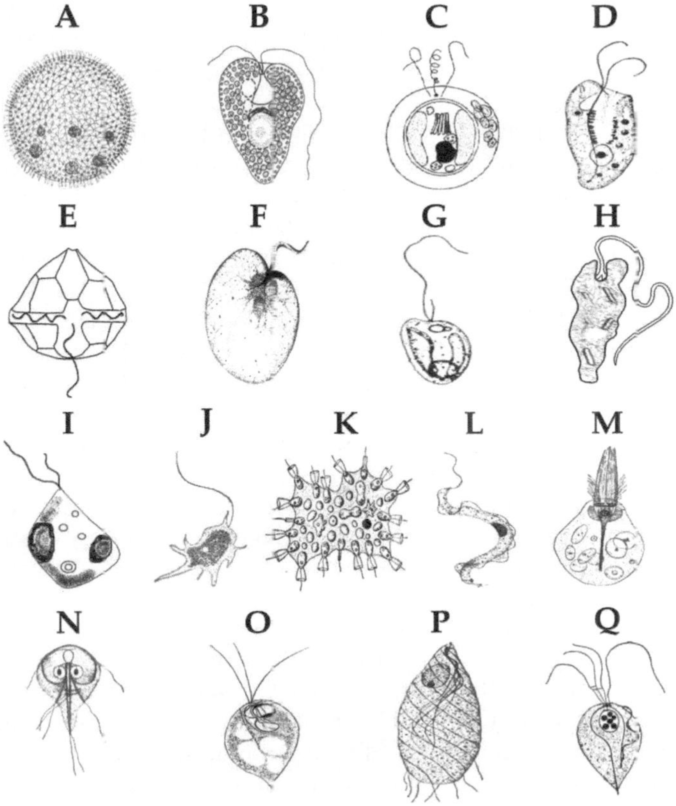

FIGURE 1.3

Mastigophora: (A) Volvocida: *Volvox* (after Janet, 1922), (B) Chloromonadida: *Vacuolaria viridis* (from Menezes and Bicudo, 2010) (C) Coccolithophorida: *Coccolithus* (*paleonerdish.wordpress.com*), (D) Cryptomonadida: *Chilomonas* (after Prowozek, 1903), (E) Dinoflagellida: thecate *Peridinium* (redrawn from Salmaso and Toletti, 2009), (F) *Noctiluca scintillans* (*oceandatacenter.uscs.edu*), (G) Chrysomonadida: *Ochromonas* (after Conrad, 1926), (H) Euglenida: *Euglena* (after Baker, 1933), (I) Heterochlorida: *Heterochloris mutabilis* (*ftpmirror.your.org*), (J) Rhizomastigida: *Mastigamoeba* (after Calkins, 1901), (K) Choanoflagellida: *Proterospongia* (after Kent, 1881), (L) Bisosecida: *Trypanosoma remaki* (after Minchin, 1909), (M) Hypermastigida: *Lophomonas* (after Kudo, 1926), (N) Diplomonodida: *Giardia enterica* (after Kofoid and Swezy, 1919), (O) Retortamonadida: *Chilomastix mesnili*, (P) Oxymonadida: *Pyrsonympha* (after Powell, 1928), (Q) Trichomonadida: *Trichomonas buccalis* (after Hinshaw, 1926).

insects (Hyman, 1940). (vi) Among dinoflagellates, the family Symbiodiniaceae comprise several genera (e.g. *Symbiodinium*), species and subspecies (Hill et al., 2019) and are symbionts of the coral-building cnidarians. (A) On the negative side, a few dinoflagellates generate toxic blooms and red tides (e.g. *Gonyaulux catenella*); the red-tide producing *Protogonyaulux* secretes poison that causes paralysis in shellfish. (B) The parasitic *Ichthyodinium chabelardi* destroys eggs of sardines and (C) *Blastodinium* castrates its copepod host (Finlay and Esteban, 2018).

Perhaps for the first time, the number and distribution pattern of flagellates are looked at from the angle of species diversity. As already indicated, most flagellates have two flagella, although the number varies from one to many. Exceptionally, *Paramastix* has two rows of 8–12 flagella (see Warren et al., 2016). The 125 speciose Choanoflagellida have one long flagellum surrounded by a transparent collar (Fig. 1.3K). The Bisosoecida resemble choanoflagellates in containing one flagellum but lack the collar. In the 76 (23 *Trypanosoma* sp + 53 *Leishmania* sp [*Wikipedia*]) speciose suborder Trypanosomatina, the single flagellum forms the border of its wavy extension of the plasma membrane called undulating membrane (Fig. 1.3L). The 100 speciose Cryptomonadida have a pair of flagella similar in size and function (Fig. 1.3D). The Chrysomonadida, with a limited number of species, have one long and another shorter flagellum (Fig. 1.3G). Contrastingly, the 1,000 speciose Euglenida also have two flagella, but one longer (mostly 50–100 µm in length; unusually 500 µm in *Euglena oxyuris*, Hyman, 1940) is backwardly directed trailing flagellum (usually in contact with substratum, Sanders, 2009) to anchor and steer the euglenoid, while the other (usually hidden behind the euglenoids) propels the animal forward (Finlay and Esteban, 2018). Similarly, the 2,000 speciose Dinoflagellida are characterized by two unequal flagella (Fig. 1.3E), one spiraling transversely around the body axis in its cingulum or girdle slow spinning motion, when swimming (Sanders, 2009) and the other longitudinal to the axis and distally directed from its sulcus. *These two groups make up 43% of all Mastigophora. The dissimilar flagella with different functions in a sort of 'division of labor' in Euglenida and Dinoflagellida have driven their evolution toward more and more species diversity.*

Rhizopoda are Protozoa, in which pseudopodia serve as the only means of locomotion and food capture during the whole or part (encystation) of the life cycle. They are much less organized than flagellates. In many species, the cycle, however, includes the production of flagellate swarms (e.g. *Paramoeba*) and some are flagellates as adults and thereby evince their close relationship to flagellates. Based on the structure of pseudopodial organization, the superclass Rhizopoda is divided into two classes namely Rhizopodea and Actinopodea. The former comprises four subclasses and > 30 orders, and the latter five subclasses (Table 1.3). The Rhizopodea are characterized by naked Amoebida (e.g. *Amoeba dubia*, Fig. 1.4A), shelled Arcellinida (e.g. *Difflugia*, Fig. 1.4D), tapering filopods in *Filosia* (e.g. *Gromia*, Fig. 1.4F) and reticulated rhizopods in the subclass Granuloreticulosia including the order Foraminiferida (e.g. *Rotalia*, Fig. 1.4G). The class Actinopodea is characterized by radiating axopods and includes those with central capsule perforated, as in Radiolaria (e.g. *Thalassicola*, Fig. 1.4J) or non-perforated, as in Acantharia (e.g. *Acanthometra*, Fig. 1.4K), or without the capsule, as in Heliozoa (e.g. *Actinosphaerium*, Fig. 1.4L). The other subclasses consist of parasites of algae in Proteomyxidia (e.g. *Vampyrella*, Fig. 1.4N) or in animals Piroplasmea (e.g. *Babesia canis*, Fig. 1.4O).

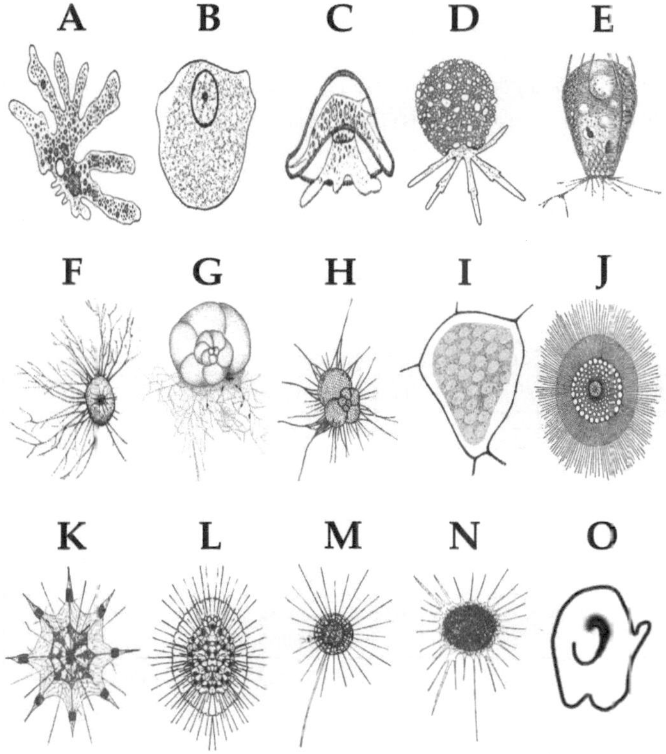

FIGURE 1.4

Rhizopoda: (A) Amoebida: *Amoeba dubia*, (B) *Entamoeba histolytica* (after Swezy, 1920), Arcellinida: (C) *Arcella*, (D) *Difflugia*, (E) *Euglypha* (after Leidy, 1879), (F) Filosia: *Gromia* (after Jepps, 1926), (G) Foraminiferida: *Rotalia* (from Finlay and Esteban, 2018), (H) *Globigerina* (free hand drawings from Hyman, 1940), (I) Mycetozoia: *Plasmodiophora* (free hand drawing), (J) Radiolaria: *Thalassicola* (after Huth, 1913), (K) Acantharia: *Acanthometra* (after Moroff and Stasny, 1909), (L) Heliozoa: *Actinosphaerium*, (M) *Clathrulina* (free hand drawing from Hyman, 1940), (N) Proteomyxidia: *Vampyrella* (free hand drawings from Hyman, 1940), (O) Piroplasmea: *Babesia canis* (after Nuttall and Smith, 1907).

Firstly, of 11,550 rhizopods, ~ 8,700 species inclusive of 4,500 speciose foraminifers and 4,200 speciose radiolarians (Table 1.3) are marine inhabitants. Incidentally, Gnanamuthu (1943) commenced the studies on foraminifera collected from the famous Krusadai Island, Gulf of Mannar, South India. In their informative monograph, Murugesan et al. (2021) have recorded the existence of 383 species from the South Indian Coast. Secondly, among the three classes that engage one or other locomotory organelle, the rhizopods are the slowest and least efficient motiles (Hyman, 1940). Hence, the need to escape from their predators is obvious. Approximately 6,500 rhizopod species belonging to the groups arcellinids (2,000 species) and foraminifers (4,500 species) are fortified by a variety of testate covering or shell. Briefly, *the species diversity in rhizopods seems to have been driven in the following descending order: marine habitat > shelled protection.*

TABLE 1.3

Superclass (pseudopodial) 11,550 speciose Rhizopoda (from Corliss, 2001*, Kosakyan et al., 2020[†], Adlard and O'Donoghue, 1998[‡], *onezoom.org*)

Taxon	Name	Characteristics	Species (no.)
Class 1	Rhizopodea	Lobopodian or filopodian pseudopod	7000
Subclass 1	Lobosia	Lobopodian pseudopod	
Order 1	Amoebida	Naked, lobose pseudopodia	
Order 2	Arcellinida	Shelled – 2 vesicular nuclei	2000[†]
Subclass 2	Filosia	Tapering, branching filopodia	300
Subclass 3	Granuloreticulosia	Reticulate rhizopodia	
Order 3	Foraminiferida	Chambered – Branching pseudopodia	4500*
Subclass 4	Mycetozoia	Trophic – amoeboid develop into multi-nucleate plasmodium	
Class 2	Actinopodea	Axopodia radiating from spherical body	4550
Subclass 5	Radiolaria	Marine – perforated central capsule	4200
Subclass 6	Acantharia	Marine – non-perforated capsule	
Subclass 7	Heliozoa	No central capsule – radiating filopodium	100*
Subclass 8	Proteomyxidia	Parasitic on plants/algae	250[‡]
Subclass 9	Piroplasmea	Parasitic in blood cells of vertebrates	

Some biologically interesting rhizopods should be mentioned (i) Though foraminifers are marine habitants, a few of them that form simple oval tectinous shell (e.g. *Gromia, Allogromia* and *Lieberkuhnia*) occur in freshwater. (ii) The heliozoan helioflagellids are flagellated. (iii) A group of rhizopods with radiating axopods are amoeboid as parasites (e.g. *Babesia*, Fig. 1.4O) but flagellated, when free-living (Fig. 1.4N). (iv) The foraminifers capture their prey using the reticulate network, but the axopods paralyze (by toxic substance) their prey, when touched an axopod. (v) In the testate amoebae, the prey size is limited by shell aperture size (see Fig. 1.4D, E, Hyman, 1940). (vi) Some free-living anaerobic amoebae (e.g. *Mastigamoeba, Mastigella, Pelomyxa*) do not have mitochondria; these pelebionts support several types of endosymbiotic bacteria including methanogens. (vii) Planktonic foraminifers (e.g. *Globorotalia ruber*) engage algal symbionts that are protected from digestion. In benthic foraminifers, the symbiont diversity reaches a climax including cyanobiont: *Nitzchia,* dinoflagellate: *Symbiodinium microadriatum* and chlorophyte: *Chlamydomonas*. Their numerical diversity can be several thousand/10 cm^2 area in the Mississippi Delta (Finlay and Esteban, 2018). (viii) More interestingly, the fossil remains of the planktonic foraminifers can be used to estimate the sea surface palaeotemperature. About 90% of their shell is composed of calcium carbonate, which is crystallized and mineralized as calcite (Hyman, 1940). Sea water and calcite differ in their $^{13}\delta$:$^{18}\delta$ ratio. With increasing ratio, the temperature also increases. Hence, the $^{13}\delta$:$^{18}\delta$ ratio is used to estimate the sea surface temperature of the geological past (Finlay and Esteban, 2018). (ix) The usefulness of the foraminifers, as sensitive indicators in petroleum exploration and sea level rise during the geological past has already been described (see p 7).

The superclass Sporozoa is classified into following five classes, as listed in Table 1.4. A few representative examples are shown in Fig. 1.5.

TABLE 1.4

Classification of superclass 6,500 speciose Sporozoa. Value arrived by subtracting species for all other groups 5,788 from 6,500 (*parasite.org.au*)

Taxon	Name	Characteristics	Species (no.)
Class 1	Telosporea	Gliders – spores without polar capsule	
Subclass 1	Gregarinia	Large extracellular trophozoites – sexual reproduction only, e.g. *Lecudina, Seleridium, Stylo-cephalus, Schizocystis*	1650 (*Wikipedia*)
Subclass 2	Coccidia	Intracellular – schizogonic clonal multiplication, e.g. *Aggregata*	712
Order 1	Eucoccida		
Suborder A	Eimeriina	Macrogametocyte produce many microgametes, e.g. *Eimeria*	1530 (Levine, 1980)
Suborder B	Haemosporina	Macro- and micro-gametes develop independently, e.g. *Plasmodium*	156 (*cdc.gov*)
Class 2	Microsporidea	Small spores of unicellular origin – parasites of invertebrates and vertebrates, e.g. *Amblyospora*	1300 (Cali et al., 2017)
Class 3	Haplosporea	Spores – schizogony, e.g. *Haplosporidium*	51 (*WoRMS*)
Class 4	Toxoplasmea	No spores – Binary fission – naked sporozoites, e.g. *Toxoplasma*	1 (Woodhall et al., 2014)
Class 5	Myxosporidea	Large spores of multicellular origin – fish parasites, e.g. *Myxobolus*	1,100 (Lom, 1984)

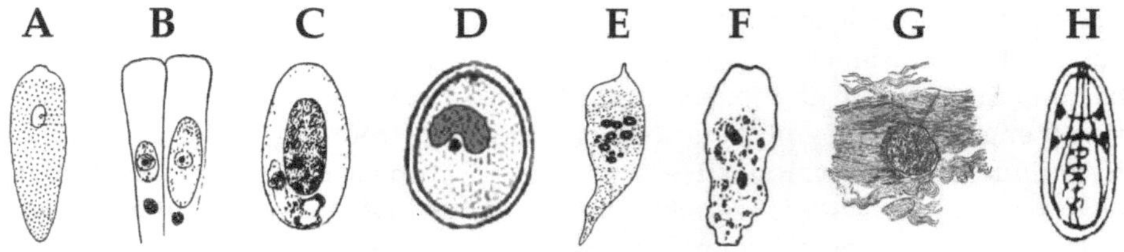

FIGURE 1.5

Sporozoa: (A) Extracellular Gregarinia: *Gregarina garnhami*, (B) Intracellular Eimeriina: *Eimeria* from centipede intestinal cells (from D.H. Wenrich), (C) Haemosporina: intracellular *Plasmodium* (after C. Huff), (D) Haplosporea: intracellular *Ichthyosporidium*, (E) Myxosporidea: *Ceratomyxa* (trophozoite), (F) *Myxidium* (trophozoite), (G) *Myxobolus* (within fish muscle), (H) *Nosema bombycis* spore (after Stempell, 1909).

Sporozoa comprise exclusively inter- or intra-cellular parasitic Protozoa on invertebrate and vertebrate hosts. Their classification is based on the type of life cycle. The ensuing description is based on a simple explanatory statement by Hyman (1940). The sporozoan parasites are transmitted from one host to the other mainly through hard walled spores, which sporulate into infective mostly crescentic or less often spherical, oval or elongated or worm-like sporozoites. On successful entry, they may remain motionless or exhibit amoeboid, euglenoid or gliding movements, as in gregarines, which have myonemes and special adhesive organelles. The growing sporozoites called trophozoites, on attaining maturation size, directly clonally multiply and become agamonts known as schizonts and schizogonic cycle is repeated to produce merozoites, which transform into sporozoites and auto-infect more and more new host cells. However, a few sporozoites develop into gamonts, which directly generate gametes; following fertilization, the zygote may encyst and enter into sporogony, as in *Eimeria*. Alternatively, some trophozoites transform into gametocytes to generate anisogamic gametes, as in *Plasmodium*. Following fertilization, the zygotes are injected by a sanguivorous host, in which sporogony occurs, i.e. the cyst formation is avoided.

A few interesting reports on Sporozoa may be listed: (i) The sporozoans are considered as host-specific. However, the following observations in some species may question it. For example, (a) *Haemogregarina bigemina* is reported from 60 host fish species from the North and South Atlantic, South Pacific, Mediterranean and Red Seas. (b) *Myxidium incurvatum* is described from 20–30 host fish species (in different orders) inhabiting geographically diverse zones (Lom, 1984). (c) Monniot (1990) lists 45 coccidian species including 35 in the genus *Lankesteria*; of them, *L. amaroucii* infect at least six ascidian species. (d) Many species of *Aggregata* are host specific to their cephalopod hosts but not to the decapod crabs (Hochberg, 1990). (e) *Toxoplasma gondii* infects many felids, the Definitive Hosts (DHs) and many herbivorous birds and others, which serve as Intermediate Hosts (IHs) (see Woodhall et al., 2014).

(ii) A series of other reports shows that *if not host specific, many sporozoans are sex specific with regard to their hosts.* (a) An unidentified gregarine is known to infect only females of the black lipped oyster *Crassostrea echinata* (Lauckner, 1983). This is also true for the gregarine *Merocystis tellinovum* infecting the bivalve *Tellina tenuis* (Lauckner, 1983). (b) Regarding *Eimeria brevoortiana*, the merogonic and gamogonic cycles take place in the pyloric caecal epithelium of the menhaden *Brevoortia tyrannus*. Its zygote actively migrates to the testis, in which sporogony is completed. Understandably, *E. brevoortiana* can be successful only in male menhaden (see Lom, 1984). (c) On the other hand, many sporozoans are female specific. It is well known that *Plasmodium* spp are reported only from the sanguivorous female (*Anopheles*) mosquito. (d) Many microsporideans are also female-specific. *Thelohania herediteria* infect the ovaries and muscles of females but not males of *Gammarus duebeni*. *Octosporea effeminans* also infect only the ovaries and adipose tissues of *G. duebeni* (Meyers, 1990). Overstreet and Weidner (1974) describe sterilization of female hosts by *Indosporous octospora* in a few *Palaemon* host species – *P. rectirostris* and *P. serratus* in the French Coast, *P. serratus* from the coast of England and *P. elegans* in the Black Sea, Romania. (e) More interestingly, *Amblyospora* spp infect only females of mosquito larva + adult as well as copepods (see Fig. 1.19). It may not be a surprise, if all the 1,300 Microsporidea are female specific. In that case, along with the 156 speciose Haemosporina, some 23% of all sporozoans may prove to be female host-specific.

(iii) Sporozoan incidence is rare in Platyhelminthes. But a careful observation by Overstreet (1978) showed that *Urosporidium crescens* is a hyperparasite on the microphallid metacercaria *Microcephalus basodactylophallus* encysted in the blue crab. (iv) Heavy infection by *Eimeria sardinae* can cause serious lesions on *Sardina sprattus*. *Leostomus xanthurus* may suffer from tumor on infection by *Ichthyosporidium gigantea*. The mature oocytes of the coccid *Goussia gadi* continuously fall into the swim bladder of the cod *Gadus morhua*, haddock *Melanogrammus aeglefinus* and pollock *Pollachuis virens* until their bladders are crammed up and become functionless. The microsporid *Glugea hertwigi* infect gonads of the European and American salmonids *Osmerus eperlanus* and *O. mordax*, respectively, block the gonadal opening and thereby renders the discharge of eggs and sperms impossible. Eventually, *G. hertwigi* leads to the death of its hosts. In Canada alone, for example, the loss can be several million tons of fishes. Sporadic events have been reported for mass mortality of 70 tons of dead salmonid smelts washed ashore in a 120 km stretch. *Myxobolus exiguus* is known as a rather harmless parasite on *Mugil auratus* and *M. cephalus*. However, the gills of these mullets can be so fully packed with *My. exiguus* cyst rendering the gills functionless and causing death due to asphyxia and bleeding. In a sporadic event, the dead mullets were washed ashore in the Black Sea and Azov Sea in quantities of 600 kg/km (Lom, 1984).

Ciliophora, the most homogenous and monophyletic group, are distinguished by the cilia as locomotor and food-capturing organelles and intriguing dual nuclear dimorphism. In many ways, they are the most specialized protozoans with (i) elaborate food-capturing ciliature and (ii) pellicular differentiation, (iii) development of myonemes and fibrillar elements as well as neurofibrils to integrate these specialized organelles. The cilia are short filaments that appear from basal bodies in the ectoplasm and pierce the pellicle to the exterior. The ciliature is arranged in an orderly row, called kinetics and the cilia in these rows beat with effective waving strokes – all in the same direction (Horan, 2003). The arrangement is an important morphological feature for identification and classification. The superclass Ciliophora is divided into two major classes (Table 1.5). The class Opalinata has oblique rows of cilia, no cytostome but one homomorphic nucleus and are mostly parasites. The class Euciliata includes four subclasses: The euciliates are characterized by the presence of simple or compound ciliature and (ii) permanent cytostome and cytopyge. 1. The subclass Holotricha consists of seven orders. 2. The subclass Peritricha: their body is covered by simple and uniform ciliature except for the absence of buccal cilia. They are recognized by their inverted funnel or bell-shaped bodies, which are mounted on a stalk. Their adults are devoid of cilia but have apical and buccal cilia. The third subclass Suctoria consists of sessile and stalked adults with no cilia. However, they do possess cilia during early life in the ciliated swarm stage, which enable their dispersal. The fourth subclass Spirotricha is characterized by a flattened body with locomotory cilia present only in the lower surface. The cilia are well developed and wind clockwise to the cytostome (see Horan, 2003). Figure 1.6 shows some representative species of Ciliophora.

TABLE 1.5

Classification of Superclass 8,000 speciose Ciliophora (Lynn, 2010a).* *onezoom.org*

Taxon	Name	Characteristics	Species (no.)
Class 1	Opalinata	Oblique rows of cilia – no cytostome → 2 monomorphic nucleus – parasites	200 (*Wikipedia*)
Class 2	Euciliata	Cytostome – cytotype – transverse fission	7800
Subclass 1	Holotricha	Simple uniform ciliature – no buccal cilia	
Subclass 2	Peritricha	Apical buccal cilia – no cilia in adults	1000*
Subclass 3	Suctoria	Stalked and sessile	480*
Subclass 4	Spirotrichia	Reduced somatic cilia – buccal cilia	680 (*WoRMS*)

Some interesting ciliates may be mentioned. *Chromidina elegans* is one of the largest ciliates and measures 5 mm in size (Hochberg, 1990). In the absence of food, the green *Paramecium bursaria* filled with zoochlorellae survive and multiply. On removal of micronucleus, *Euplotes* cease fission and die.

However, a few micronucleus-less mutants endure its life and divide but are incapable of conjugation and endomixis. *Paramecium* and *Stentor* distinguish nutrients and feed them but reject carmine and other toxic dyes. The *Stentor* do not recognize dead from live prey but select a few prey species among many. *Bursaria* accept a yolk particle but reject non-nutrients (Hyman, 1940). Endemism is characteristic of, perhaps, as many as 30% of ciliate species (Lynn, 2010b). As mentioned earlier, anaerobic symbiotic ciliates inhabit the

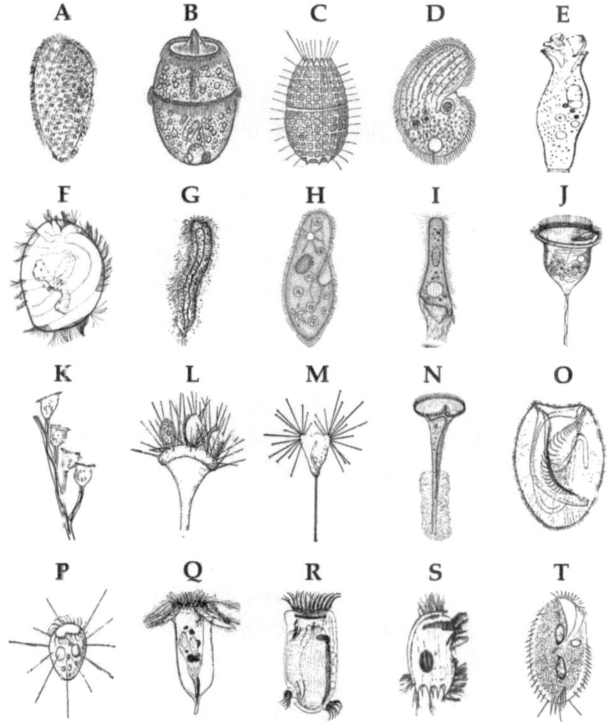

FIGURE 1.6

Ciliophora: (A) Opalinata: *Opalina*, (B) Gymnostomatida: *Didinium* (after Blochmann, 1895), (C) *Coleps* (after Conn, 1905), (D) Trichostomatida: *Colpoda* (after Rhumbler, 1888), (E) Chonotrichida: *Spirochona gemmipara* (after Hertwig, 1877), (F) Apostomatida: *Hyalophysa*, (G) Astomatida: *Anoplophrya* (after Pierantoni, 1909), (H) Hymanostomatida: *Paramecium*, (I) Thigmotrichida: *Boveria*, (J) Peritrichida: *Vorticella*, (K) *Carchesium*, (L) Suctorida: *Ephelota gigantea* (after Noble, 1929), (M) *Acineta* (after Kent, 1881), (N) Heterotrichida: *Stentor roeseli*, (O) *Bursaria* (after Brauer, 1886), (P) Oligotrichida: *Halteria* (after Kent, 1881), (Q) Tintinnida: *Favella* (Campbell, 1926), (R) Entodinomorphida: *Cycloposthium* (after Strelkow, 1929). (S) Odontostomatida: *Saprodinium*, (T) Hypotrichida: *Stylonychia*. A, F, H, I, J, K, N, S and T are redrawn from different sources.

rumen of ruminants and caecum of other mammals with post-gastric fermentation (Finlay and Esteban, 2018). The parasitic rhynchodid *Sphenophyra* spp are considered as host specific. But of 14 species, *S. dosiniae* can infect at least five bivalve species (Lauckner, 1983). *Orchitophyra stellarum* is parasitic on starfishes. Interestingly, *O. stellarum* infect 13% female but 1% male *Asterias forbesi*. In an extreme case of *A. rubens*, only males are infected (Jangoux, 1990).

1.3 Life Cycles

As they provide the base to build all other aspects of diversity, life cycles of protozoans must be described. In Protozoa, sexual reproduction is rarely associated with increase in the number of individuals. It interrupts clonal multiplication and rapid population growth by multiple binary fission. However, it is essential to prevent senescence and extinction of clonal lineages (Sanders, 2009). In free-living and parasitic protozoans, life cycles are unusually diverse and greatly complicated, especially with the involvement of two hosts (e.g. *Toxoplasma gondii*, Fig. 1.21). These complications have limited the description of the cycle to only a few species in any taxonomic group. *For example, the cycle is described only for 30 species or 0.67% of the 4,500 speciose free-living foraminifers (Goldstein, 1999). For parasitic protozoans, the description is limited to 6 species or 0.54% of the 1,100 speciose parasitic Myxosporidea* (Noble, 1944). *These values may be compared with 3% for free-living polychaetes* (see Pandian, 2019).

Before describing the life cycle types, another preamble must to be included. (i) For the first time, the cycle is viewed from the angle of ploidy status in different stages of gamogony, sporogony and schizogony, where it is included. Only a few publications indicate the location of meiosis or mitosis but most others do not even hint at them, especially in a few rhizopods. From scattered relevant information collected from different sources, an attempt is made to locate mitosis and/or meiosis at appropriate locations in all the cycle types. However, some of which may have to be confirmed. The emerging new picture seems to narrate the trial and tribulations undergone by sporozoan parasitic groups in their struggle to get transmitted from one host to the other.

It may be appropriate here to define a few terms also (see Noble, 1944).

Gamogony is the process, by which one or more gametes are produced from a gamete-mother cell, the gamont or gametocyte.

Sporogony is the process, by which spores or sporozoites are developed from a zygote or sporoblast (e.g. Myxosporidea).

Schizogony is the process of binary or multiple fissions during clonal phase of trophozoites, and results in the production of "merozoites".

Nucleogony replaces schizogony in Myxosporidea.

Plasmogamy (or hologamy) is the cytoplasmic union of two or more cells. The nuclei of the cells remain distinct.

Trophozoite is the growing vegetative individual.

Schizont is the trophozoite during the process of schizogony.

Sporont (sporoblast) is an individual or stage, which is destined to develop into one or more spores, either directly by the secretion of a sporocyst membrane or indirectly by internal development resulting in one or more spores (e.g. Myxosporidea).

Pansporoblast is a sporont, which produces two or more spores.

Sporoplasm is the vital living germ of the spores, as distinguished from the other differentiated parts. It includes the sporoplasm nulcei, which are the gametes.

Sporulation is the formation of spores by multiple divisions. This process is called sporogony, when it follows sexual fusion.

Synkaryon is the nucleus of a zygote.

Plasmotomy is the division of multinucleate body into smaller, separate bodies. This cytoplasmic division is independent of nuclear division.

Endodyogony is the expanded version of sporogony in Toxoplasmea.

1.3.1 Mastigophora

The 6,900 speciose Mastigophora are a fascinating taxonomic group, as they are the ancestors of all other protozoans. They include ~ 4,845 speciose photosynthetic autotrophic plant-like Phytomastigophora and ~ 2,055 speciose heterotrophic animal-like Zoomastigophora. Within the former, some like the euglenids and dinoflagellates can be autotrophic, mixotrophic or heterotrophic and are included as algae, as well (Guiry, 2012). The haplontic algal life cycle is dominated by the relatively longer lasting gametophytic or reproductive phase and is also characterized by mitotically generated gametes and meiosis succeeding fertilization, as against the diploid angiospermic plants and metazoans with meiotically generated haploid gametes and meiosis preceding fertilization (Pandian, 2022). Hence, the identification of ploidy level in mastigophores becomes of paramount importance.

The life cycles of Mastigophora are diverse and complicated except in free-living Euglenida and parasitic zoomastigophores. Repeated computer searches with key words 'life cycles in Mastigophora' and 'life cycles in flagellates' have yielded only scattered information. Hyman (1940) also hints of only limited information. Being widely diverse, there seems to be no common unifying life cycle for Mastigophora. Hence, the superclass must be considered as polyphyletic assemblage. Yet, definitive and/or putative descriptions for their life cycles could be inferred for the < 100 speciose Cryptomonadida (Fig. 1.7A–B), possibly < 100 speciose Chrysomonadida, < 100 speciose Coccolithophora (Fig. 1.7C–D), ~ 400 speciose Volvocida (Fig. 1.8A), 2,000 speciose Dinoflagellida and 1,000 speciose Euglenida among the ~ 4,845 speciose Phytomastigophora (Tables 1.2, 1.6). And the 125 speciose free-living Chaonoflagellida, ~ 100 speciose two hosted Trypanosomatina (Fig. 1.8B) and single hosted 1,800 species included in five orders of the other symbiotic or parasitic Zoomastigophora. On the whole, relevant information for the cycle could be assembled for 5,615 species or 81% of the flagellates.

TABLE 1.6

Ploidy status in sexual and clonal reproduction in taxonomic groups of Mastigophora (Hyman, 1940*). (2n) and (n) indicates the occurrence of clonal reproductive cycle

Order	Sexual	Encystment	Clonal	Reference
		Phytomastigophorea (~ 4,845 species)		
Cryptomonadida	n → 2n+n → n	Cyst*	n	*site.iugaza.edu.ps*
Chrysomonadina	n → n+n → n	Cyst*	n	Hyman (1940)
Coccolithophorida	n → (2n) → meiosis → (n) ← fusion ↙			Eikrem et al. (2017)
Chloromonadida			n? cyst	cf Lackey, J.B.
Volvocida	n → 2n → n	Cyst*	n	Umen (2020)
Dinoflagellida	n → 2n → n	Cyst*	n	Salmaso and Tolotti (2009)
Euglenida	Not occur*	Cyst	n	*geochembio.com*
Heterochlorida	Unknown			
		Zoomastigophora (~ 2,055 species)		
Rhizomastigida	Unknown			
Choanoflagellida	(2n) → n → 2n → meiosis → (n) ← fusion ↙			Levin and King (2013)
Kinetoplastida	Unknown			
Bisosoecida			2n	Peacock et al. (2014)
Hypermastigida	Not occur*		2n? cyst	Hyman (1940)
Diplomonadida	Not occur*		2n? cyst	*ndvsu.org*
Retortamonadida	Not occur*		2n? cyst	*cdc.gov*
Oxymonadida	Not occur*		2n? cyst	Hyman (1940)
Trichomonadida	Not occur*		2n?	*slideshare.net*

Notably, the classification of flagellates is not based on the types of life cycle, as in Sporozoa. Still, the following seven life cycle types can be recognized, and related to the taxonomic groups. Of the seven types, four are from Phytomastigophora and remaining three from Zoomastigophora. Notably, this identification of life cycle is based on only one or two representative example species.

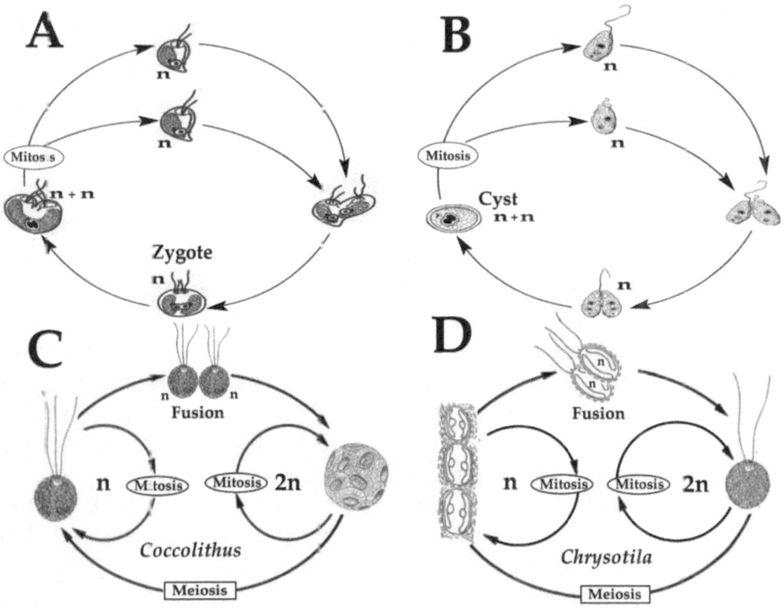

FIGURE 1.7

Life cycles in mastigophoran (A) isogamous hologamy in *Cryptomonas* (based on *site.iugaza.edu.ps*) and (B) anisogamous hologamy in *Chilomonas* (based on Hyman, 1940). (C) *Coccolithus* and (D) *Chrysotila* (modified from Eikrem et al., 2017).

1. The most ancestral flagellates are perhaps the cryptomonadids and possibly chrysomonadids. They do not have distinct mitotically generated either haploid or diploid gametes. They simply engage the 'transformed' vegetative stage as 'iso'- (e.g. *Cryptomonas*, Fig. 1.7A) or 'aniso'- (e.g. *Chilomonas*, Fig. 1.7B) 'gametes' in hologamic syzygy, which results in naked, as in the former or encysted, as in the latter with n + n zygote. The former subitaneously undergoes mitosis to produce two haploid young ones; the latter does the same, but after a period of dormancy. Incidentally, hologamy involves the adherence of two mature individuals in a sort of syzygy state and their nuclei stick together (see Fig. 17J of Hyman, 1940). However, cytological evidence for their fusion is not yet available. Hyman (1940) chose to name the hologamy as a sexual cycle. *The hologamy seems to be the first discovery to include sexual reproduction, besides clonal multiplication.*

2. The second in the series are the coccolithophores. They are also haplontics and display one of the most complicated life cycles. In them also, no gametes are generated, instead they *engage the 'transformed' vegetative stage as isogamous gametes* (Fig. 1.7 C–D). They undertake clonal multiplicative cycles independently as haploid and diploid. In another cycle, some of these clonal 'vegetative' haploids, following isogamous fusion, reenter into the cycle (Table 1.6). These diploids undergo meiosis and the resulting haploids enter into a clonal cycle. In these three groups of flagellates, the non-sexualized reproduction has reduced the scope for generation of new gene combinations. Thereby, their diversity is limited to < 100 species. 3. Incidentally, the life cycle of 125 speciose choanoflagellates is also as complicated as that of coccolithophores. However, they are sexualized and pass through 2n → n → 2n sequence in sexual cycle (Table 1.6).

The third in the series namely *the volvocids are the first to discover the distinct mitotically generated anisogamic gametes* (Fig. 1.8A). Following oogamous (sperms reach the immotile oocyte) fertilization, the diploid zygote is formed within a cyst. After a period of dormancy, the zygote undergoes meiosis and releases haploid flagellated primordial young ones. With further mitoses, the young ones pass through multicellular and gonadium stages and eventually establish a haploid colony. *The volvocids and dinoflagellates share a common life cycle and pass-through n → 2n → n ploidy status in sexual cycle* (Table 1.6). *They clonally multiply as haploids alone.* The solitary haploid chlamydomonads mitotically generate haploid motile zoospores or mating types, which can be distinguished functionally but not morphologically (Fig. 5.1A). They are 'heterogametics' as haploids but not as diploids, as in animals and some dioecious angiosperm (Coelho et al., 2019, Umen and Coelho, 2019). 4. The fourth in the series are the euglenoids. Typical of the haplontic algae, they are haploids. A few of them are autotrophs/mixotrophs but majority of them are heterotrophs. They undergo repeated clonal multiplications as haploids. For example, *Entosiphon sulcatum* pass through 947 generations, as haploid clonal "without losing vitality and no change in fission rate"; they also "indicate no change in the periodic nuclear reorganization process" (Hyman, 1940). Euglenoids lack the sexual reproductive cycle, except perhaps in *Scytomonas*, in which hologamy has been reported (see Hyman, 1940).

5. The fifth in the series are the choanoflagellids. Their sexual life cycle, as represented by *Salpingoeca rosetta* (Levin and King, 2013), which pass through 2n → n → 2n ploidy status during sexual reproduction. Their diploids generate anisogamic but flagellated motile zoospores. Following fertilization, the zygote meiotically generates young ones (see Levin and King, 2013). 6. The sixth life cycle includes that of trypanosomatina, in which clonal multiplication alone occurs, as diploids in both of their hosts (Fig. 1.8 B), i.e. clonal schizogony in human or livestock, and clonal sporogony in tsetse fly or dog flea or sand flay (for the flagella-less *Leishmania*). Notably, experimental investigation has shown that they are diploid at all stages in their cycle (Peacock et al., 2014). 7. The remaining five orders of

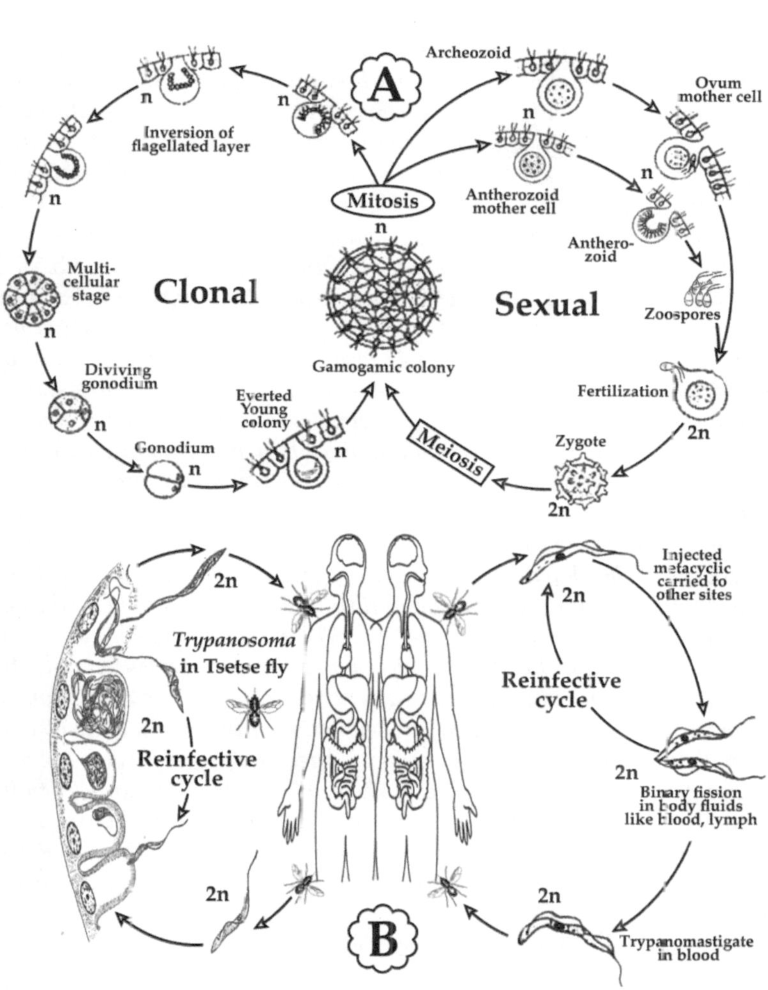

FIGURE 1.8

(A) Clonal and sexual life cycles in a flagellate colony of *Volvox*. (B) Clonal life cycle of *Trypanosoma/Leishmania* in tsetse fly *Glossina*/dog flea *Ctenocephalides*/sandfly *Phlebotomus* and in human and livestock (compiled and simplified from Hyman, 1940 and others).

Zoomastigophora are single hosted symbionts or parasites and are diploids. They lack a sexual cycle, as euglenids do. Most of them are encysted, when released through the host's feces into water. On the whole, for the first time, this incisive analysis has brought to light the following new findings: 1. *The alga-like Phytomastigophora are haploids. In their sexual cycle, they pass through n → 2n → n ploidy status. Contrastingly, the animal-like Zoomastigophora are diploids and pass through 2n → n → 2n ploidy status or as diploids in clonal multiplication* (Table 1.6). *Hence, the transition from alga-like haploid status to animal-like diploid status is a landmark event in the evolutionary history of protozoans and the diploid status is conserved in most other protozoans and metazoans, as well.* 2A. *Remarkably, the sexual cycle has been secondarily lost* (cf Hawes, 1963) *in the 1,000 speciose free-living phytomastigophoran euglenids and 1,800 speciose parasites in the five orders of Zoomastigophora.* 2B. The *sexual cycle is also lost in the < 100 speciose trypanosomatina. On the whole, some 2,900 species or 42% flagellates have secondarily lost the sexual cycle* (Table 5.1). 3. *The sexual reproductive potential is low in most free-living flagellates. For example, two progenies appear from two parents in hologamic cryptomonadids and possibly chrysomonadids, as well. Dinoflagellates also generate only a maximum of four progeny/sexual cycle* (*e.g. Gloeodinium montanum,* Kelley and Pfiester, 1990). *This holds also true for the others. For example, the parasitic flagellates, in which the sexual cycle is secondarily lost, clonally produce four progeny/2n cyst in Giardia. Even among the trypanosomes, not more than a dozen sporozoites are generated/ sporogonic cycle* (see Fig. 32, Hyman, 1940). *Still, no precise information is yet available on the number of clonal cycles prior to the interruption by the sexual cycle in Mastigophora. The clonal cycle of Euglena gracilis requires 1.1 d/cycle* (Wang et al., 2018) (Table 1.7). *In comparison to rhizopods and ciliates, the food capturing efficiency of phagotrophic flagellates is less. The flagellates prefer to invest their hard-earned resources on clonal multiplication rather than sexual reproduction, as indicated by the fewer number of offspring generated by sexual reproduction, as well as the relatively longer time required for completion of the sexual cycle* (Table 1.7). 4. *The clonal multiplication seems to generate adequate progenies at a faster rate in all parasitic Zoomastigophora to sustain them. Hence, there is no need to include schizogony in their clonal cycle except the re-infective phase in trypanosomas. Or the schizogony is not yet discovered by the flagellates. 5. The seven different reproductive cycles clearly indicate that Mastigophora are rather an assemblage than monophyletic.*

Another analysis reveals that the lowest efficient mechanism of food capturing has limited species diversity to the lowest number of 6,900 species among protozoans. Consequently, *the 4,845 speciose Phytomastigophora or 70% of Mastigophora depend on autotrophy either partially or totally. Within Phytomastigophora, both dinoflagellates and euglenids have explored their dependence not only to autotrophy but also other forms of food acquisition like heterotrophy.* For example, a few dinoflagellates are autotrophs (e.g. *Cyclotella*), while others are osmotrophs, mixotrophs (e.g. *Ceratium furca*) or heterotrophs (e.g. *Gymnodinium helveticum,* see Ismael, 2003). This is also true of the Zoomastigophora. They have explored coloniality in some Choanoflagellida (17 species, Table 1.12), symbiosis (263 species in insects, see Table 4.3) and parasitism (2,900 species, Table 1.2). Briefly, *evolution in Mastigophora seems to have driven them to retain autotrophy in 70% flagellates and in the other 30% more and more toward symbiosis or parasitism.*

TABLE 1.7

Multiplication rate in sexual and clonal reproduction in taxonomic groups Mastigophora

Order	Sexual	Clonal	Doubling time	Reference
	Progeny (no.)			
Phytomastigophora				
Cryptomonadida	2/2 parent	2/parent	< 1 day	Salmaso and Tolotti (2009)
Chrysomonadida	2/2 parent	2/parent	< 1 day	
Volvocida	Many/parent*	4–6/parent*		Hyman (1940)*
Dinoflagellida	4/parent	> 4/parent	2–30 day	Salmaso and Tolotti (2009)
Euglenida	–	32–64/parent	< 1.1 day	*geochembio.com*, Wang et al. (2018)
Choanoflagellida	–	–	~ 6 h in *Monosiga brevicollis*	King et al. (2009)
Bisosoecida	–	> 12 sporozite/ cycle	–	Hyman (1940, see Fig. 32)
Hypermastigida	–	4/cyst cycle	–	
Diplomonadida	–	4/cyst cycle	9–12 h	*medscape.com*
Retortamonodida	–	4/cyst cycle	–	–
Oxymonadida	–	4/cyst cycle	–	–
Trichomonadida	–	4/cycle	~ 2.3 h	Paget and Lloyd (1990)

1.3.2 Rhizopoda

Descriptions for the life cycle of Rhizopoda remain incomplete (e.g. Radiolaria) or continuously changing. Irrespective of it, seven life cycle types can be recognized among the 11,500 speciose rhizopods. Type 1 **Hologamy**: As in Mastigophora, the arcellinids, represented by *Difflugia urceolata*, undertake hologamic 'sexual' cycle. In them, one of the compatible two individuals leaves its shell and enters into the other's (Fig. 1.9A). As in cryptomonadids, the two individuals hang on together in a sort of syzygy state. No evidence is available on their nuclear fusion. Subsequently, only two progenies emerge from the cyst. There are succinct differences between the cryptomonadids and arcellinids. Diploid hologamy occurs within a cyst in arcellinids but encystation is not obligatory for the haploid cryptomonadids. In general, the arcellinids multiply mostly clonally by binary fission (Fig. 1.9A). One of its progenies gains the shell, but the other, who receives the required material, builds its own. Type 2 **Autogamy** commences with encystation of a diploid individual in Heliozoa (Fig. 1.9B); within the cyst, it undergoes two successive antecedent meioses to generate two 'viable gametes' and six degenerating polar bodies; the event is similar to that in gametogenesis in Metazoa. Followed by isogamic (or anisogamic, see Fig. 44F of Hyman, 1940) fusion, a diploid zygote is formed. Eventually, two diploid progenies emerge from the cyst through the binary fission. Briefly, autogamy may facilitate gaining new gene combinations through meiosis. However, the scope for generation of new combination is limited, as the progenies appear from a single parent – a situation parallel to that in hermaphrodites. Type 3 **Conjugation**: The filosian *Gromia* spp are perhaps the first to generate

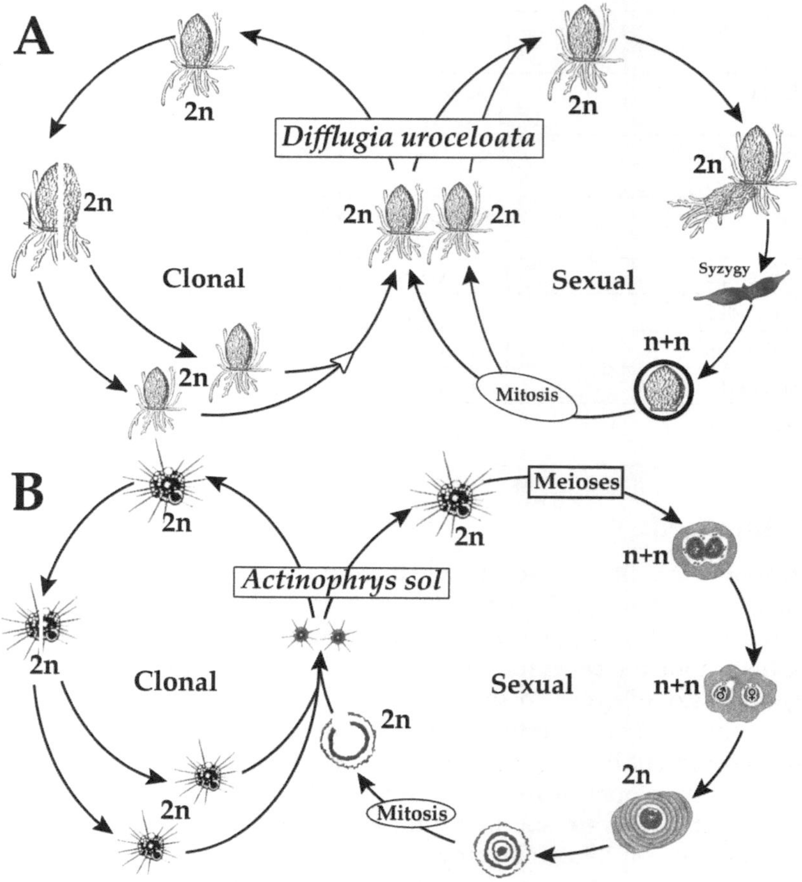

FIGURE 1.9

Life cycle of (A) the arcellinid *Difflugia urceolata* (based on Borradaile et al., 1977, Yang and Shen, 2005) and (B) heliozoan *Actinophrys sol* (redrawn from Gardiner, 1972, Borradaile et al., 1977).

biflagellated gametes, probably by mitosis. During conjugation, the conjugants exchange their gametes and produce many diploid amoeboid swarms that ultimately complete the cycle following the formation of an organic test around them (see Arnold, 1966, Psi Wave Function, February, 2011). The filosians may be the forerunners of the ciliate that sexually reproduce by conjugation. However, conjugation in the Filosia is limited to exchange of diploid gametes between the conjugants and evidence for their fusion is wanted.

4. **Amoeboid** type consists of three subtypes. Its subtype 1 includes free-living amoebae. Their cycle is exclusively clonal. A large number of daughter progenies are released from the encysted cycle, whereas only two progenies are produced by a binary fission in their subitaneous cycle (Fig. 1.10A). With 500–600 chromosomes, *Amoeba proteus* is possibly a polyploid (see Hawes, 1963), which has eliminated the scope for sexual reproduction. Recently, Naiyer et al. (2019) reported the presence of meiotic genes in *Entamoeba histolytica* cyst. Surprisingly, the members in the second subtype consisting of parasitic amoebae do not include clonal multiplication in their life cycle (Fig. 1.10B). The third Myxamoeba subtype is an interesting social group and unique to include the sexual cycle, besides clonal cycle both as individual and social group (Fig. 1.11). In *Dictyostelium discoideum*, for example, two

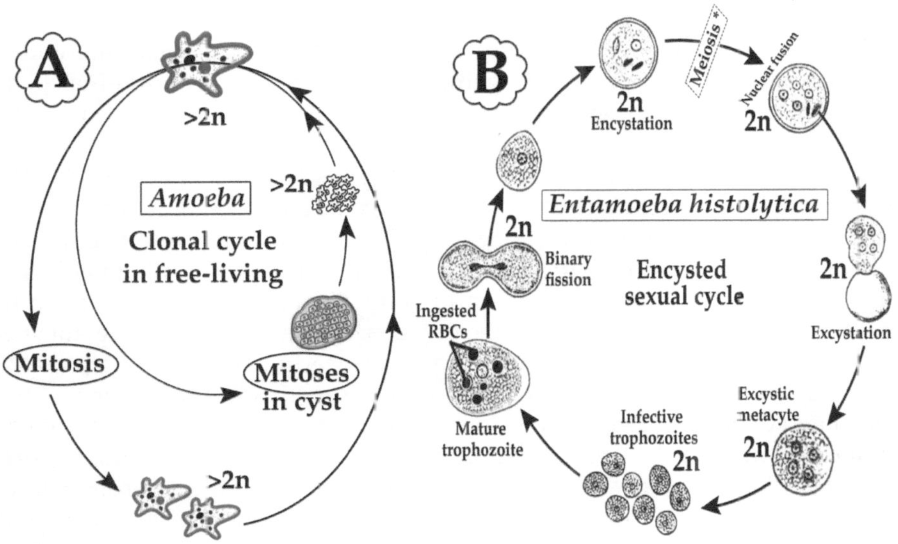

FIGURE 1.10

Life history of the (A) free-living *Amoeba* with clonal cycle alone (based on Hawes, 1963) and (B) parasitic *Entamoeba histolytica* with sexual cycle alone (based on Naiyer et al., 2019).

compatible individual fuse and the 'progeny' cannibalistically ingests the neighboring individuals. Subsequent to its enlargement and encystation, the diploid undergoes meiosis + mitosis within the cyst, from which ~ 8 diploid progenies emerge.

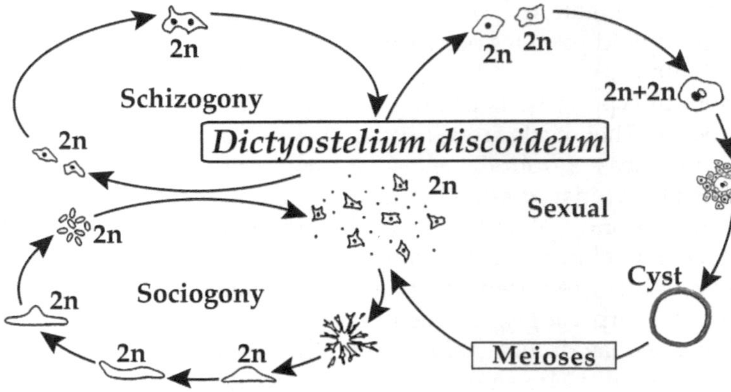

FIGURE 1.11

Life history of the myxamoeboid *Dictyostelium discoideum* with inclusions of sexual, sporogonic and sociogonic cycles (redrawn from Flowers et al., 2010).

5. **Foran** type. As in foraminifer *Elphidium*, the sexual cycle alternates with sporogony and involves isogametes generated by an individual that fuse only with those of another individual. The sporogonic cycle includes schizogony also (Fig. 1.12). *Thereby, foraminifers are the first to discover alternation of generation and the inclusion of proper schizogony in the clonal cycle of protozoans*. The cycle includes mostly alternation of sexual macrospheric shell morph with clonal microspheric shell morph, i.e. the interruption of sexual reproduction following every clonal multiplication. This may have necessitated schizogony, besides sporogony. The cycle commences from the haploid gamont, which mitotically generate eight isogonic gametes. Interestingly, *Glabratella sulcata* ensures cross fertilization. Though risky, it can ensure relatively a greater number of new gene combinations (K.G. Grell cited in Hawes, 1963). On fusion, the diploid agamont is formed, which undergoes mitoses to produce 12 haploid schizonts, each of which pass through schizogonic cycle prior to entry as diploid agamont. From a few partially known life history of some 26–28 foraminiferan species, Goldstein (1999) reported that (i) the alternation between sexual and clonal cycles is obligatory in 57% species, most probably necessary in another 21% species but facultative in the remaining 22% species. About 54% species generate bi-flagellate gametes, ~ 12% tri-flagellate gametes and 35% amoeboid gametes.

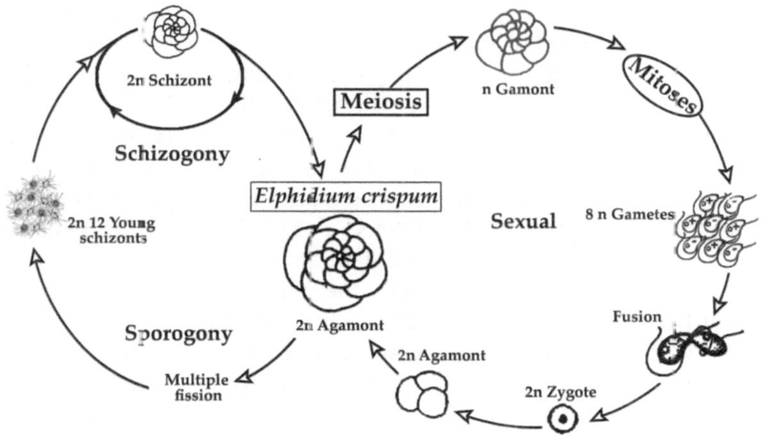

FIGURE 1.12

Life cycle in the foraminiferan *Elphidium crispum* (compiled and redrawn from K.G. Grell cited by Hawes, 1963, Hyman, 1940, Goldstein, 1999).

6. **Radiolaria** type: For 4,200 speciose radiolarians, the life cycle is not known completely. Perhaps the most complete cycle is reported for phaeodarian *Anlacantha scolymantha*, albeit evidence for the sexual cycle is scarcer due to its small size and sinking of its encysted gamont to benthic zone. In *Cercomonas longicauda*, the swarmers are known to fuse and form a zygote, which following mitoses, may produce up to 100 nuclei. From this scattered information, a tentative life cycle for radiolarians is illustrated in Fig. 1.13.

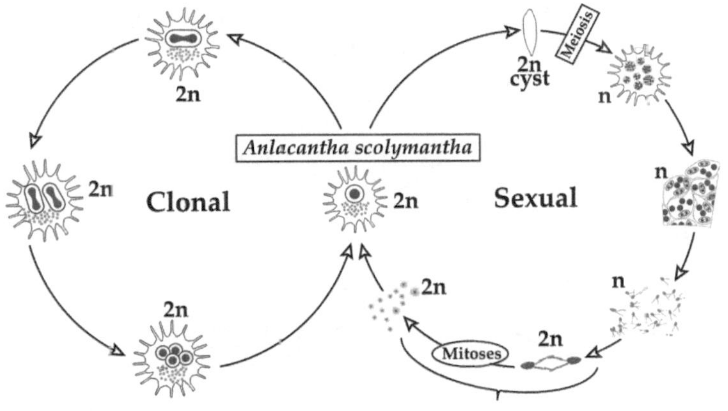

FIGURE 1.13

Life cycles of phaeodarian radiolarian *Anlacantha scolymantha* (compiled and redrawn from Lahr et al., 2011, Decelle et al., 2013, Suzuki and Not, 2015 and others). Note the marked portion of life cycle is not so far described.

7. **Parasitic** type: The superclass Rhizopoda comprises two subclasses namely the single-hosted Proteomyxidia parasitic on plants and the two-hosted Piroplasmea. The complete life cycle for the latter is described for *Plasmodiophora brassicae,* a parasite in the root gall of cabbage (*Brassica* sp). Its cycle includes clonal and sexual cycles – all within the gall (Fig. 1.14B). It is transmitted from one host to another as a 2n zygote. The rhizopodan Piroplasmea may have to be distinguished from the five speciose eucoccidian haemogregarines – parasite of fish (see p 35). The proteomyxidians are corpuscular parasites in the intermediate host (IH), the horse and intestinal parasites in the definitive host (DH), the ixodid tick. The parasites are transmitted from horse → tick → horse, when the tick bites the horse (Fig. 1.14A). Many details, which vary for species belonging to the genera *Babesia* and *Theileria,* are described by Jalovecka et al. (2018).

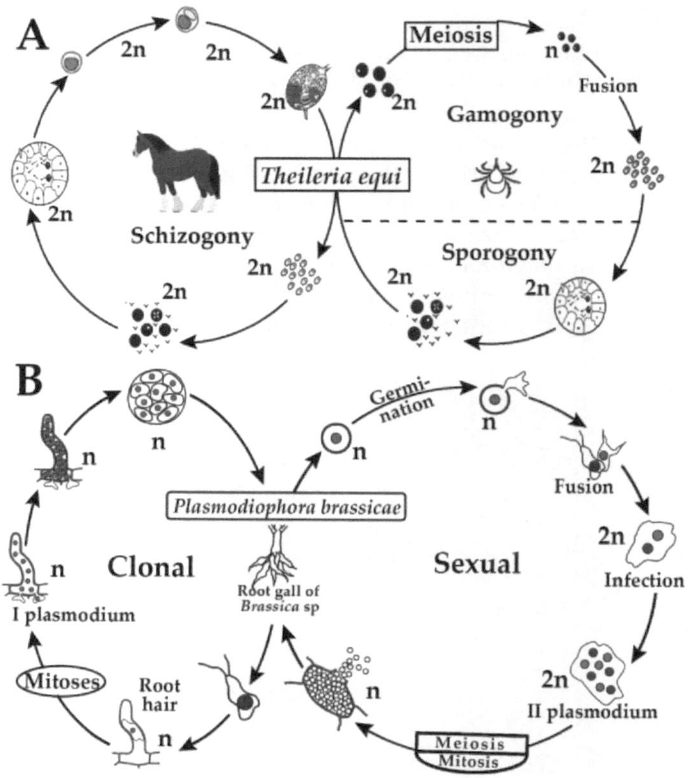

FIGURE 1.14

Life cycles (A) in proteomyxidian *Theileria equi* (based on Jalovecka et al., 2018) and (B) piroplasmean *Plasmodiophora brassicae* (modified from M. Piepenbring, *wikipedia*).

For the first time, this incisive analysis has brought to light the following new findings: 1. The occurrence of seven life cycle types along with wide differences in the ploidy sequence in sexual and clonal cycles shows that Rhizopoda are more an assemblage rather than monophyletics. 2. Rhizopoda are the most speciose taxonomic groups among protozoans. 3. Of 11,550 rhizopod species, 8,700 species or 75% are constituted by the 4,500 speciose foraminifers and 4,200 speciose radiolarians. The reasons for the diversity of foraminifers, for example, are attributed to their reproductive mode but not to the presence of the shell allowing protection against predators, as the 2,000 speciose arcellinids are also protected by a shell. (a) Unlike in most other free-living protozoans, sexual and clonal cycles obligately alternate with each other; sexual cycle generates four progeny/cell (due to fusion of eight isogametes, Fig. 1.12A) and thereby provides reasonable scope for the generation of new gene combinations/reproductive cycle. (b) To compensate for the interruption of each clonal cycle, sporogony produces 12 schizonts through multiple fission and each schizont produces multiple number of agamonts through a schizogonic cycle. Briefly, *foraminifers have discovered an innovative sexual cycle to ensure adequate number of new gene combinations and generation of adequate number of progenies through sporogony + schizogony within each reproductive cycle.* 4. In rhizopods, evolution seems to have driven food capturing mechanism from the costliest lobosian mode to the costlier one in the reticulate mode in foraminifers and to the costly one in the axopod mode in free-living Actinopodea. Table 1.8 summarizes the possible ploidy sequences in sexual and clonal cycles of rhizopods. Unlike mastigophores, the sequence for rhizopods varies widely, confirming that they are polyphyletics.

TABLE 1.8

Ploidy status in sexual and clonal reproduction in taxonomic groups of Rhizopoda. Hyman (1940)*

Order	Sexual	Encystment	Clonal
Rhizopodea			
Lobosia			
Amoebida			
Free-living	Not known	Present/absent	> 2n
Parasitic	2n → n → 2n	Present	Not known
Social	2n + 2n → 2n	Present/absent	2n
Arcellinida	2n → 2n+2n → 2 n	Present	2n
Filosia	2n → 2n → 2n	Unknown	2n
Foraminiferida	2n → n → 2n	Absent	2n
Actinopodea			
Radiolaria	2n → n → 2n	Present	2n
Heliozoa	2n → 2n+2n → 2n	Present	2n
Proteomyxidia	2n + n → 2n	Present	2n
Piroplasmea	n → 2n → n	Present	n

1.3.3 Sporozoa

They comprise 6,500 species, inclusive of 1,650 for Gregarinia, 712 for Coccidia and 1,530 for eucoccidian Eimeriina (within which 80 species are marine haemogregarines + 5 species for piroplasms, Lom, 1984), 1,300 for Microsporidea, 156 for Haemosporina, 1 for Toxoplasmea and 1,100 for Myxosporidea (Table 1.4). In the 1,650 speciose Gregarinia, the complete life cycle is described for five species; of them three and one each are distributed in marine, freshwater and terrestrial habitats. Hence, a compromise was arrived at 990, 330 and 330 for the said habitats. The 712 speciose Coccidea, represented by *Aggregata eberthi*, are assigned to marine habitats, although Hochberg (1990) listed only 25 identified and unidentified species. Of them, some 1,445 out of 1,530 eucoccidean Eimeriinia are considered as terrestrial inhabitants, after subtracting the marine 85 haemogregarine and piroplasm species. The 51 speciose Haplosporea are all marine inhabitants (Lauckner, 1983). So are the 1,100 species Myxosporidea, albeit a few like *Myxobolus* and *Henneguya* infect fresh water fish (Hyman, 1940). Among the 1,300 speciose Microsporidea, many are marine inhabitants (e.g. *Ameson, Indosporus, Microsporidium, Octosporea, Pleistophora, Thelohania*, Hyman, 1940), a few are from freshwaters (e.g. *Glugea, Mrazakia*, Hyman, 1940) and very few (e.g. *Nosema*) are from terrestrial habitat. Within *Nosema, N. bombycis*, causing pebrine disease in silkworm is terrestrial but some like *N. algerae* (Undeen et al., 1984) are aquatic. On the whole, a sum of 3,710 species or 57%, 718 or 11%, 156 or 2% and 1,915 or 30% sporozoans live in marine, freshwater, freshwater cum terrestrial and terrestrial habitat, respectively (Table 1.9). A look at the limited number of described sporozoan life cycles suggests how the following 10 types are accommodated in the classification of the superclass.

Class 1 Telosporea comprise gliders, whose spores have no polar capsule and filament. Typically, the invasive stages are crescent shaped sporozoites, which have a well-defined anterior – posterior axis (Sibley et al., 1998) sporozoites.

Subclass 1: The 1,650 speciose Gregarinia are characterized by large extracellular trophozoites and trophic transmission. In them, Group 1 consists of the apicomplexine gregarines (e.g. *Nematopsis ostrearum*, Fig. 1.16A). In them, the cycle involves (i) gamogony and sporogony alone, (ii) in two invertebrate hosts with production of 2n gymnospore from the decapod host (Fig. 1.16A) and (iii) encystation in the mussel host. In group 2, as exemplified by *Lecudina* (Fig. 1.15A) and *Schizocystis* (Fig. 1.15B) syzygy is followed by meiosis and mitosis to generate a large number of anisogametes and their fusion. However, contrasting differences between *Lecudina* and *Schizocystis* are listed:

	Lecudina	*Schizocystis*
Gamogony	In host	In water
Sporogony	In aquatic sediment	In host
Ploidy	Diploid sporogony	Haploid schizogony
Schizogony	Not included	Included

Subclass 2: Coccidia are characterized by small intracellular trophozoites and inclusion of schizogony.

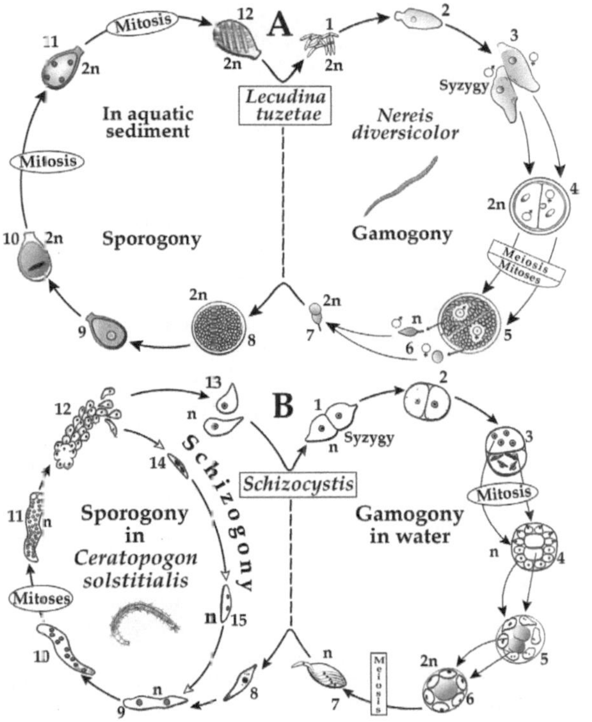

FIGURE 1.15

Life cycles of gregarines: (A) *Lecudina tuzetae*: Stages 1 to 3 transformation of sporozoites into trophozoites, 4 to 7 gamogony: syzygy → encysted gamont → meiosis/mitoses → cellularized gamont → anisogamic gametes → zygote in polychaete → encysted sporoblast → mitoses → 8 sporozoites/cyst. (B) *Schizocystis gregarinoides*: Stages 1 to 7: gamogony: syzygy → encysted gamont with peripheral pear-shaped eight male + one spherical female gametes → fertilization → meiosis → mitoses in water → 12 cysts with sporoblasts, each containing 8 sporozoites in aquatic dipteran larva (redrawn from Schrevel, 1969, Leger, 1909). In all the stages, ploidy status requires confirmation.

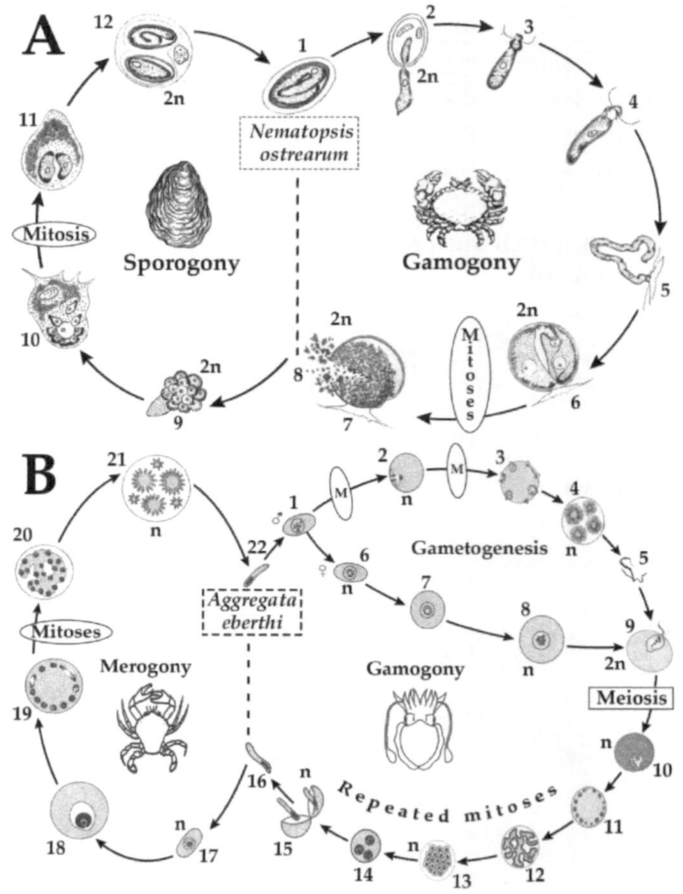

FIGURE 1.16

Life cycles of (A) gregarine *Nematopsis ostrearum*: In crab, gamogony occurs. Stages 1 to 5: transformation of sporozoite into trophozoite, which follows syzygy → mitoses → gymnospores. On rupture, released sporoblast into water is engulfed by oyster → sporogony in oyster's phagocytes → encystation → sporulation by binary fission → two sporozoite/cyst. (B) Coccid *Aggregata eberthi*. Stages 1 to 5: merozoite → trophozoite → two successive mitoses → four biflagellated sperms. Stages 6 to 9: one sporozoite → one immotile egg → fertilization → zygote → mitoses in stages 11 to 14. Sporoblast → sporozoites from *Sepia officianalis* released into water. Infective sporozoites undergo merogony → each sporocyst contain three sporozoites in crab *Portunus depurator*. Definite stage, in which meiosis occurs, is marked. Appropriate stage, in which mitosis is likely to occur, is also marked (redrawn from Prytherci, 1940, Hochberg, 1983).

Order 1: Eucoccidia are known by the intracellular trophozoites on the intestinal epithelial or blood cells, as in *Aggregata eberthi* (Fig. 1.16B). Interestingly, the number of sporozoids/sproblast per se is increased from three in *A. eberthi* on the highly motile *Sepia officianalis* (Fig. 1.16B) to six in *A. kudoi* on *S. elliptica* and to 8–16 and 12–28 in *A. octopiana* and *A. spinosa*, respectively on the relatively less motile *Octopus vulgaris* (see Hochberg, 1990).

Suborder A: Eimeriina are characterized by non-motile zygotes and production of sporozoites in sporocyst, e.g. *Eimeria* (Fig. 1.17). In them, the number of diploid schizogony cycles is fixed, species-specific and usually three or four (Stevens, 1998). In *Eimeria*, haploid gamogonic and diploid schizogonic cycles occur in the host chick but encysted oocytes on defecation are dispersed in soil. Each cyst has four sporocysts, each with two sporozoite/cyst (see Lom, 1984).

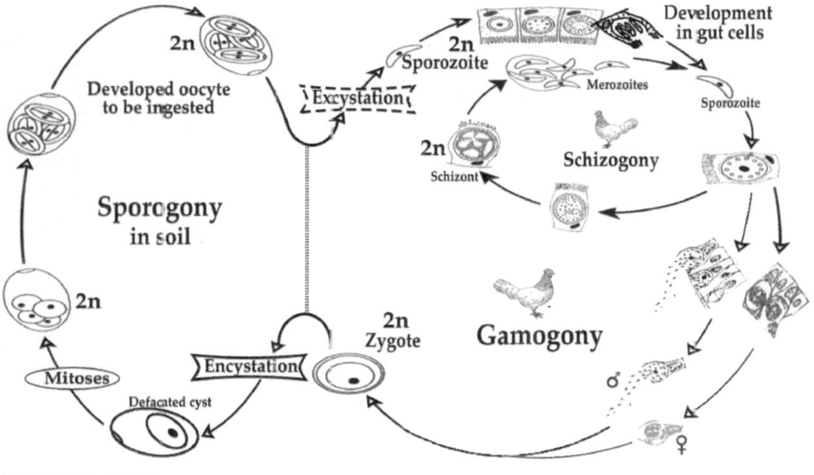

FIGURE 1.17

Life cycle of *Eimeria*. Note both schizogony and gamogony take place in the host chick (compiled from different sources).

In another suborder Sepatatina (Lauckner, 1983), the presence of extra-intestinal intracellular 80 speciose haemogregarines and five speciose piroplasms is reported (Lom, 1984). They are parasites of marine fish and include 80 speciose haemogregarines and five speciose piroplasms. Whereas the former infects the leucocytes or white blood corpuscles (RBCs) in schizogony and erythrocytes or Red Blood Corpuscles (RBCs) in gamogony, the latter only depend on RBCs. In schizogony, the crescentic sporozoites undergo divisions into 2/RBC in *Haemogregarina bigemina*, 4/RBC in *H. quadrigemina* from *Callionymus lyra*), 8/RBC in *H. simondi* or 16/RBC in

Gobius paganellus. A few vermicular merozoites leave the RBCs and generate anisogamic gametes in the blood plasma, where they fuse to form a zygote. They are found as gametes, zygotes, oocysts and sporozoites, in the haemocoelom of Intermediate [invertebrate] Hosts (IHs) like leeches (e.g. *Hemibdella soleae*), isopods (e.g. *Gnathia maxillaris*) or copepods (e.g. *Lernaeocera*). Rarely, gamogony for *Cyrilla uncinata* takes place in the leech *Johanssonia* sp (Lom, 1984). For piroplasms, Khan (1980) reported some interesting information. They may not be host specific, for the inoculation of the infected blood from its natural host *Lycodes lavalaei* or *L. vahlii* into perciformids *Myoxocephalus octodecemspinosus* and *Anarhichus lupus* is followed by further development; However, it fails to develop in the pleuronectid *Pseudopleuronectes americanus* or the gadid *Gadus morhua*. The natural infection of *Haemosporidium beckeri* is known from the leech *Platybdella olriki*. Interestingly, *L. lavalaei*, fed on the leech, is successfully infected. Hence, the transmission from DH to leech and IH to DH. But is not known how the IH in isopod and copepod acquire the parasite.

Suborder B: Haemosporina are characterized by a motile-zygote and a naked sporozoite. In *Plasmodium* (Fig. 1.18), the expanded hepatic and erythrocytic schizogonic cycles take place in vertebrate host, and gamogony and sporogony in mosquitos belong to the genus *Anopheles*. *Plasmodium* spp cause the devastating malarial disease in man and his livestock; their life cycle is elaborately described in Chapter 4. The complete life cycle of human malarial parasite *Plasmodium* in man and *Anopheles* was discovered and

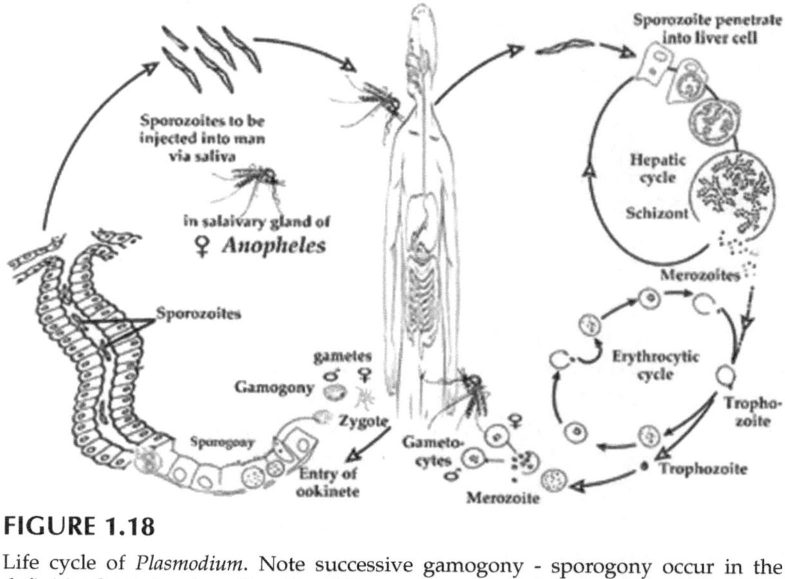

FIGURE 1.18

Life cycle of *Plasmodium*. Note successive gamogony - sporogony occur in the definitive host *Anopheles* ♀ and hepatic - erythrocytic cycles in man (compiled from Simonetti, 1996 and others).

described by Sir Ronald Ross from India, for which he was awarded the Nobel Prize in 1902.

Class 2: Microsporidea are cytotoxic intracellular parasites of arthropods and vertebrates. Their small spores are of unicellular origin. Approximately, nine species/y were erected from 2006 to 2017 (1,200 species, Didier and Weiss, 2006, 1,300 species, Cali et al., 2017). In their review, Cali et al. (2017) described extracellular/intracellular infective stages under three major groups: (i) *Vittaforma*, (ii) *Brachiola ichthyosporidium* and *Nosema* (iii) *Thelohania*. The trophic or autoinfective spores, sporoplasmic and sporogonic phases are described for a few species like *Endoreticulatus*, *Enterocystozoon*, *Glugea*, *Pleistophora* and *Trachipleistophora*. However, a more complete cycle is pictorially represented for *Amblyospora* (Fig. 1.21). The cycle involves schizogony in a female adult mosquito → n binucleate spores → transovarial transmission to its female F_1 larva → following meiosis spore generates n meiospores → trophic transmission → female copepod → sporogony → release of uninucleate n spores into water → entry into filter-feeding mosquito larva → schizogonic multiplication of trophozoites.

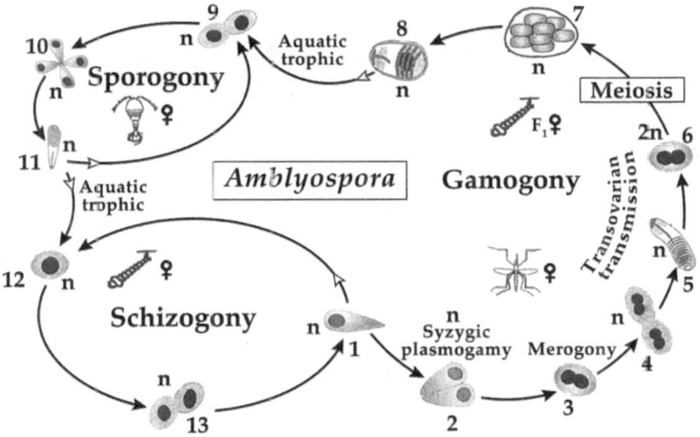

FIGURE 1.19

Life cycles: Microsporidean *Amblyospora*. Schizogony from stage 12 → 13 → 1 → 12 in ♀ mosquito larva. Extended gamogony in stages 1 → 6 via transovarian transmission to adult. Sporogony in stages 9 → 10 → 11 → 9 occurs in ♀ copepod (modified from Robert-Gongneux and Darde, 2012, Andreadis, 2007).

Class 3: As described in Fig. 1.19, Haplosporea are characterized by one oyster host alone, e.g. *Haplosporidium nelsoni*.

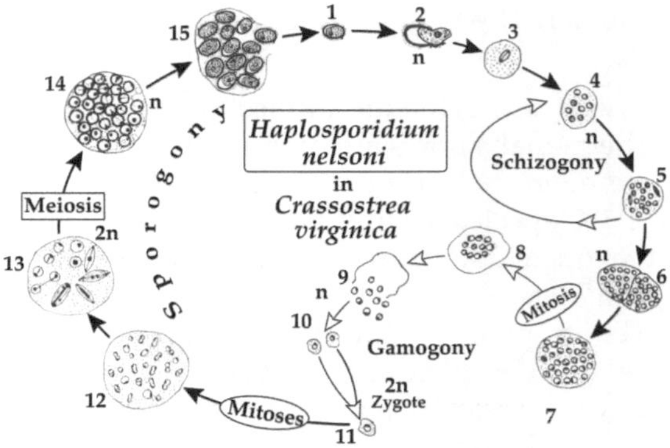

FIGURE 1.20

Proposed life cycle in *Haplosporidium nelsoni*. Schizogonic stages 4 → 5 → 4, gamogonic stages 7–11 involving isogametes and sporogony by repeated mitoses followed by meiosis in stages 12 → 15 and release of sporozoites at stage 15 (modified and simplified from Farley, 1967).

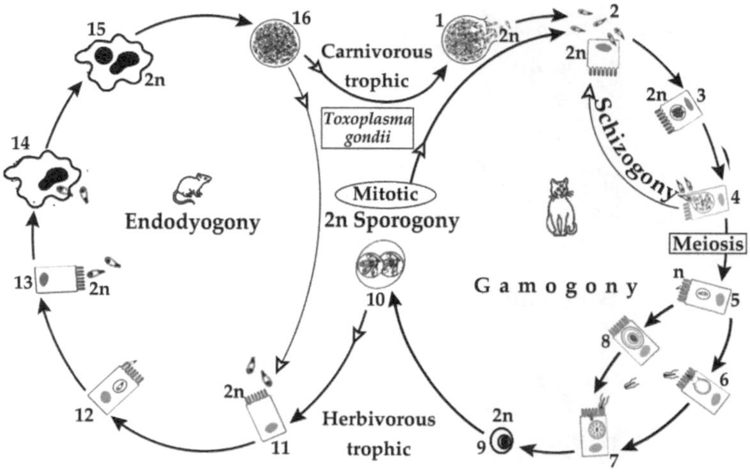

FIGURE 1.21

Life cycle of *Toxoplasma gondii*. Schizogamy in stages 2 → 4 → 2 and gamogony in stages 5 → 9 take place in cats. From shortened sporogony, either trophozoites reinfect the same cat or release it as a cyst eaten by the rat, in which endodyogony in stages 11 → 16 and schizogonic-like stage 16 → 11 → 16 takes place in the rat (modified from Robert-Gongneux and Darde, 2012, Andreadis, 2007).

Class 4: Toxoplasmea with a single species is characterized by the absence of spores, clonal multiplication only by binary fission and many sporozoites in the cyst. In the life history of *Toxoplasma gondii*, repeated schizogonic diploid cycles and haploid gamogony through anisogamy take place in the gut epithelial cell in a felid host (Fig. 1.20). A condensed sporogony occurs in the encysted sporont voided into soil. On trophic transmission to herbivorous hosts, sporozoites undergo endodyogony to propagate the parasites into all the host's organs. As if to compensate the risk involved in transmission, eight sporozoites are produced from each sporont.

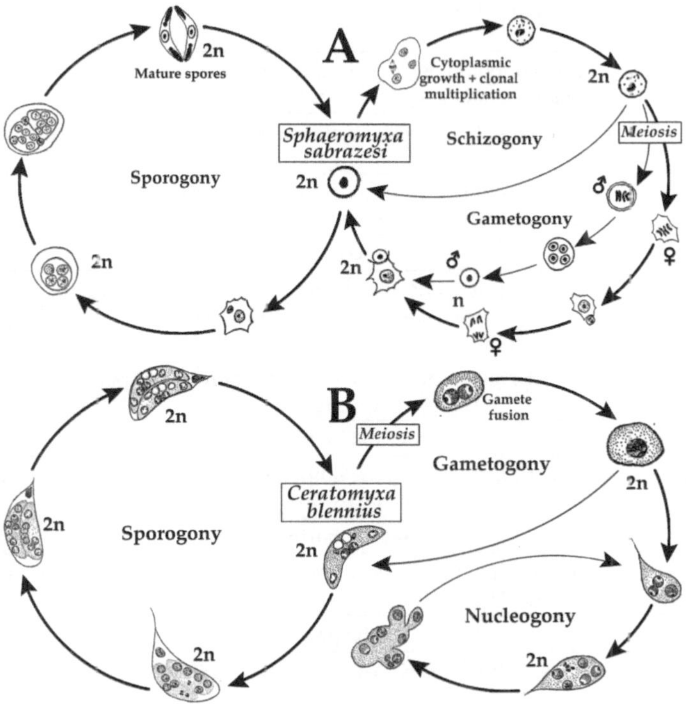

FIGURE 1.22

Life cycle of Myxosporidea: (A) *Sphaeromyxa sabrazesi* and (B) *Ceratomyxa blennius*. Note the replacement of schizogony by nucleogony in *C. blennius* (simplified and redrawn from Noble, 1944).

Class 5: Myxosporidea are all intercellular parasites of teleost fish. In them, new species were erected at the rate of 98/y. Thanks to Noble (1944), the complete life cycle is known for at least six myxosporean species. In their cycles, a typical sequence of schizogony → gametogony → sporogony occurs in *Sphaeromyxa sabrazesi* (Fig. 1.22A). But the cycle is altered with shorter gametogony, which is followed by nucleogony involving the syncytium-like budded nuclei within the growing cytoplasm and sporogony in *Ceratomyxa blennius* (Fig. 1.22B).

A comprehensive summary is provided in Table 1.9 for the reproductive phases and number of stages in ten sporozoan life cycle types. In Table 1.10,

TABLE 1.9

Reproductive phases and number of infective stages generated in some Sporozoa

Species	Reported/inferred observations
Telosporea: Gregarinia	
Lecudina tuzetae	One polychaete host; eight sporozoite/encysted sporont (Schrevel, 1969)
Selenidium pendula	One polychaete host; four sporozoite/encysted sporont (Schrevel, 1971)
Stylocephalus sp	One tenebrionid host; ruptured cyst releases monete chain of sporonts; eight sporozoite/encysted sporont (Schrevel and Desportes, 2016)
Schizocystis gregarinoides	One dipteran larval host; eight sporozoite/encysted sporont (Leger, 1909)
Nematopsis ostrearum	Two hosts; numerous naked gymnospores from gamogony crab; two sporozoite/encysted sporont from sporogonic oyster (Prytherci, 1940).
Coccidia	
Aggregata eberthi	Two hosts; gamogony → three naked sporonts → each produce 36 sporocysts; three naked sporozoite/sporocyst from *Sepia* three sporozoite/encysted sporont from crab (Hochberg, 1983)
Haemogregarina quadrigemina	Two hosts; four sporont/oocyte; eight sporozoite/sporont from fish host
Eucoccida: Eimeriinia	
Eimeria spp	One host; chick; schizogony limited to 3–4 cycles; eight sporozoite/encysted sporont (Stevens, 1998)
Haemosporina	
Plasmodium	Two hosts; sporogony generates 3,400 to 7,500 sporozoite/sporont (Simonetti, 1996); penetrative transmission
Haplosporea	
Haplosporidium nelsoni	One sessile host; 16 sporozoite/sporocyst (Farley, 1967)
Toxoplasmea	
Toxoplasma gondii	Two hosts; schizogony and gamogony in cats. Sporogont containing two sporonts each with four sporozoites; encysted sporont transmitted to reinfect the same cat or to rat; in the latter endodyogony including schizogony like cycles take place. Cyst with numerous sporozoites is generated from rat (Robert-Gangneux and Darde, 2012)
Myxosporidea	
	One fish host; Complete life cycle is described
Microsporidea	
Amblyospora	Two hosts; schizogony in ♀ mosquito larva, on whose metamorphosis gamogony is initiated in ♀ adult; on transovarian transmission to its F_1 larva, in which gamogony is completed. Eight meiospore/encysted oocyte are eaten by ♀ copepod. Naked spore is acquired by filter feeding ♀ mosquito larva (Cali et al., 2017)

available information is assembled for a number of species, their proportions in the three habitats, host number, transmission, encystment, schizogony and so on. For the 6,500 speciose Sporozoa, the number of species, for which life cycle is described may not exceed 10 types in approximately 100 species, most of which cause serious diseases to man and livestock. Hence, a few compromises had to be made to infer some generalizations, which shall be valid, although the number and proportion of species arranged in Table 1.10 may deviate but not significantly.

As transmission can be risky, costly and complicated, 4,891 species or 75% sporozoans have opted for the inclusion of only one host species in their life cycle (Table 1.10). *In the most speciose (1,650) Gregarinia, which only depend on one slow motile host species, evolution has proceeded with the highest species diversity. Contrastingly, sessility of the bivalve host species decelerates species diversity in the 51 speciose Haplosporea.* Within haplosporeans, like *Haplosporidium nelsoni, H. costale* and other undescribed eight species engage only sessile bivalves *Crassostrea virginica, Ostrea edulis, O. lurida, Mytilus californianum, Teredo bartschi, T. furcifera, T. navalis* and a few others (see Lauckner, 1983). The gregarinians include *Nematopsis, Porospora* and others; they engage mostly sessile hosts like oysters and mytilids. For example, *N. legeri* and *N. prytherchi* use one of the two host species but *N. duorari* and *N. scheideri* one of the four and six host species, respectively. Still, these uses involve mostly bivalves and crabs that inhabit littoral zone and thereby bring spatial proximity between the sessile bivalves and slow motile crabs to increase the chances of infection (Lauckner, 1983). Incidentally, the *host's motility alone need not necessarily accelerate species diversity*. For example, the life cycle of haemosporean *Plasmodium* spp involves the fast-moving human and flying mosquito as hosts. Still, they are diversified into 156 species only, as the mode of transmission is injective. Understandably, engaging the fast-moving fishes, Myxosporea are diversified into 1,100 species. Microsporidea, which engage also fast-moving invertebrates and vertebrates, have also diversified into 1,300 species. It is notable that the former has to select one among the 32,510 speciose teleosts (see Pandian, 2011) whereas the latter can recruit one among far more numbers of invertebrate and vertebrate host species and are relatively more diversified than the former. That takes us to the second aspect of number of host species available within a taxonomic host group for infection. For example, the eimeriians engage mostly birds, although stray incidences of *Epieimeria isabellae* from *Conger conger* and *Eimeria brevoortiana* from *Brevoortia tyrannus* are known (Lom, 1984). Interestingly, birds include 10,038 species (see Pandian, 2021b). Hence, Eimeriina have ample scope to select one or another bird from a relatively large number of host species.

TABLE 1.10

Estimation of the number of species in different taxonomic groups of sporozoan by the following characteristic features. M = marine, FW = freshwater, FWT = freshwater cum terrestrial, T = terrestrial, G = gamogony, S = schizogony, C = clonal, † = via substratum/medium, * = ingestion, Aniso = Anisogamic, Gymno = Gymnospores

Characteristics	____	Gregarinia	Coccidia	Eimerrina	Haemogregarina	Haemosporina	Toxoplasmea	Haplosporidea	Myxosporidea	Microsporidea	Total	
											(no.)	(%)
Habitat	M	990	712	–	90	–	–	51	1000	868	3710	**57**
	FW	330	–	–	–	–	–	–	100	288	718	**11**
	FWT	–	–	–	–	156	–	–	–	–	156	**2**
	T	330	–	1440	–	–	1	–	–	144	1915	**30**
Host	One	1320	–	1445	–	–	–	51	1100	975	4891	**75**
	Two	330	712	–	85	156	1	–	–	325	1609	**25**
Schizogony	Yes	330	–	1445	85	156	1	51	1100	1300	4468	**69**
	No	1320	712	–	–	–	–	–	–	–	2032	**31**
Sequence	S → G	–	–	1445	85	156	1	51	550	1300	5620	**86**
	G → S	330	–	–	–	–	–	–	550	–	880	**14**
Transmission	Passive†	1320	–	1445	–	–	–	51	–	1300	4116	**63**
	Active*	–	–	–	85	–	–	–	1100	–	1185	**18**
	Both	330	712	–	–	–	1	–	–	–	1043	**16**
	Injective	–	–	–	–	156	–	–	–	–	156	**2**
Encystment	Yes	1320	–	1445	–	–	–	51	1100	–	3916	**60**
	Naked	330	–	–	85	156	–	–	–	–	571	**9**
	Both	–	712	–	–	–	1	–	–	–	2013	**31**
Reproduction	Sex + C	1650	–	1445	85	156	1	51	1100	1300	5788	**89**
	C	–	712	–	–	–	–	–	?	–	712	**11**
Gamete	Aniso	1320	712	1445	85	156	1	–	550	–	4268	**66**
	Isogamic	–	–	–	–	–	–	51	550	1300	1901	**29**
	Gymno	330	–	–	–	–	–	–	?	–	330	**5**

Transmission from one host to other may involve passive trophic mode through an aquatic medium (e.g. *Schizocystis*, Fig. 1.15B) or watery sediment in a marine habitat, as in *Lecudina tuzetae* (Fig. 1.15A) or in soil substratum in terrestrial habitat, as in *Eimeria* (Fig. 1.16). It can be an active mode, when a benign host eats an infected host, as in *Nematopsis ostrearum*, in which the infected oyster is eaten by a benign crab (Fig. 1.16A). A combination of both passive and active trophic modes of infection takes place in two-hosted sporozoans. The following are some examples:

Parasite species	Active trophic host	Passive trophic host
Aggregata eberthi (Fig. 1.16B)	*Sepia officianalis*	*Portunus* crab
Toxoplasma gondii (Fig. 1.21)	Cat	Rat

A second active mode involves injection of sporozoites by a sanguivorous vector, as in haemosporinans (Fig. 1.16). It demands production and injective delivery of larger number of sporozoites than the passive or active trophics. For example, 3,400 (in *Plasmodium falciparum*) to 7,500 (in *P. cynomolgi*) sporozoites are produced in a mosquito, which delivers ~ 50 sporozoite/bite (Simonetti, 1996), in comparison to three naked/sporocyst (e.g. *Aggregata eberthi*) or 4–8 sporozoite/encysted sporont in active and passive trophics (e.g. gregarinians, Table 1.9). On the whole, 4,116 species or 63% and 1,185 or 18% sporozoans are transmitted by passive and active trophic modes, respectively (Table 1.10). Almost all the two-hosted sporozoans involve passive transmission from Definitive Host (DH) to Intermediate Host (IH) and an active one by DH (e.g. *A. eberthi*, Fig. 1.15B, *Haemogregarina quadrigemina*). They comprise 1,609 species or 25% sporozoans. In contrast to > 97% sporozoans, which engages passive or active or both modes of transmission, the injective mode is limited to 156 species only or 2% of sporozoans. Hence, *the trophic mode of transmission accelerates species diversity, whereas injective mode decelerates it.*

Schizogony: With clonal multiplication, schizogony extends infection to increased numbers of the host's cells in the gut epithelium or blood corpuscles. The number of its cycle is fixed at 3–4 in *Eimeria* (Fig. 1.17). However, this information is lacking for other taxonomic groups. In *Plasmodium*, schizogony is divided into initial hepatic and subsequent corpuscular cycles and thereby it is more widely distributed into alimentary and circulatory systems of man (Fig. 1.18). In *Toxoplasma*, it takes place in the gut cells of felid DH and a cycle similar to it also occurs in herbivorous IH, in which endodyogonic sporozoites infect almost all organs (Fig. 1.21). Its inclusion and expansion into different organs and systems may intensify pathogenicity and eventually lead to the host's death.

Except in *Schizocystis* (Fig. 1.15B), schizogony is not included in other gregarinians (Fig. 1.15A, Fig. 1.16A). This also holds true for coccids (e.g. *Aggregata eberthi*, Fig. 1.16B). On the whole, schizogony is included in 4,468 species or 69% of Sporozoa (Table 1.10). In the remaining 31%, it is not included. Incidentally, Hawes (1963) listed a few protozoan species, in which sexual reproduction is secondarily lost. Hence, it is not clear whether schizogony is also secondarily lost in gregarinians and coccids, or it was discovered by the eucoccids like the haemogregarine one billion years ago (BYA) and conserved in all other sporozoan groups (Fig. 10.4A). In general, schizogony and gamogony go together and occur in DH. It usually follows sporogony except in freshwater gregarine *Schizocystis gregarinoides*, in which it takes place in the aquatic dipteran larva (Fig. 1.15B). It is not again known, whether schizogony was shifted to precede gamogony in all other sporozoans. In the 1,100 speciose Myxosporidea, schizogony and its modified version of nucleogony occur (Fig. 1.22). More so, it is also not clear whether the extracellular parasitic gregarines (1,320 species) can ill afford the cost of including of schizogony.

As in other Protozoa, Sporozoa also undertake sexual reproduction to generate new gene combinations besides clonal multiplication in sporogonic and/or schizogonic cycles. Not surprisingly, the combination of sexual and clonal reproduction takes place in > 89% sporozoans; in the remaining, clonal multiplication alone occurs in the 712 speciose gregarines like *Nematopsis* (Fig. 1.16A), which constitute 11% of all sporozoans (Table 1.10). In sporozoans, anisogamy is more common (4,268 or 66%) than isogamy (1,901 species, 29%).

Encystation facilitates safer transmission from one host to the other and one habitat to the other, as well. It occurs in 3,916 species or 60% of all sporozoans (Table 1.10). In 571 species, transmission is achieved in a naked form, as the infected host is eaten by the second host in *Schizocystis* group and haemogregarines or an injection by Haemosporina. In the remaining 2,012 species or 31% sporozoans, transmission is achieved by a dual strategy.

Ploidy: Inferred from scattered relevant information, values for ploidy status in different life stages in gonogonic, sporogonic and schizogonic cycles in the selected taxonomic groups are assembled in Table 1.11. The inferred values require confirmation from future researches. Remarkably, the sequences, through which the gregarinians pass through reveal the trials and tribulations undergone by them to establish in appropriate hosts during the checkered history of evolution. During gamogony, 990 marine gregarine species pass through the meiotic sequence 2n → n → 2n; in contrast, the 330

TABLE 1.11

Ploidal status: estimation of the number of species in different taxonomic groups of Sporozoa (based on Figures 1.15 to 1.22)

——	Gregarinia	Coccidia	Eimerrina	Haemogregarina	Haemosporina	Toxoplasmea	Haplosporidea	Microsporidea	Myxosporidea	Total (no.)	Total (%)
					Gamogony						
2n → n → 2n	990	–	1445	85	156	1			1100	3777	**58**
n → 2n → n	330	712	–	–	–	–	51	1300	–	2393	**44**
2n	330	–	–	–	–	–	–	–	–	330	**6**
					Sporogony						
2n	990	–	1445	85	156	1	51	–	1100	3828	**59**
n	660	712	–	–	–	–	–	1300	–	2672	**41**
					Schizogony						
2n	–	–	1445	85	–	1	–	–	1100	2631	**59**
2n → n	–	–	–	–	156	–	–	–	–	156	**3**
n	330	–	–	–	–	–	51	1300	–	1681	**38**

2n → n → 2n = meiotic gametogenesis; n → 2n → n = mitotic gametogenesis. Schizogony is included only in 3,368 species or 52% of sporozoans. The eucoccid Septatina (e.g. *Aggregata*) are considered diploids, as their approximate equivalent *Trypanosoma brucei* is experimentally shown as diploid (Peacock et al., 2014) freshwater gregarinians pass through the mitotic sequence of n → 2n → n and the 330 speciose *Nematopsis* group through 2n. Strikingly, a common sequence is established in most other taxonomic groups of Sporozoa. For example, all the 1,445 speciose Eimeriina pass through only one sequence of 2n → n → 2n, 2n and 2n ploidy status during gamogony, sporogony, schizogony, respectively. Therefore, for the first time, *this analysis has brought to light that all the taxonomy groups except Gregarinia pass through one or another sequence during gamogony, sporogony and schizogony. It is likely that the gregarinians are polyphyletic, whereas all other taxonomic group of Sporozoa may be monophyletic.*

Interestingly, some protozoans have secondarily lost sex (Table 5.1). Similarly, a few sporozoan taxonomic groups seem to have secondarily returned to schyzogonic haploidy. They are represented by Haplosporea (see Fig. 1.20) and Microsporidea (Fig. 1.19).

1.3.4 Ciliophora

They are the second most speciose (8,000) taxonomic group among protozoans and are characterized by cilia and dual nuclearity. Vertebrate mucosal epithelium has 200–300 cilia/cell, but *Tetrahymena* and *Paramecium* have 750–4,000 cilia/individual (Bayless et al., 2019). Hence, clothing the entire body by cilia can be costly. Consequently, evolution in ciliates has proceeded toward restriction of ciliated area to a specific zone, e.g. oral zone in *Didinium* or combined to form cirrus or tentacle. Some hypotrichs use the cirri as walking legs in crawling (Hyman, 1940). The clonal cycle of ciliates is less frequently interrupted by conjugation and more frequently by endomixis. As a result, conjugation is a rare event in natural populations and many of them may permanently remain clonal alone (Lucchesi and Santangelo, 2004). Experimental investigations by Sonneborn (1939) and Jennings (1939) revealed a few eligibility criteria for conjugants. Firstly, members of the same clone are not eligible to conjugate. Secondly, each ciliate species consists of several mating types, each of which, in its turn, consists of mating groups. Conjugation occurs between members of the mating groups but not between the mating types. Following a few rounds of clonal multiplications, some daughter individuals are transformed into gamonts. The eligible conjugants become sticky and adhesive to each other with the mouth, where actual protoplasmic fusion occurs. In conjugation, major events occur in the following sequence: In the respective 2n micronuclei, meiosis produces 4 micronuclei each → Degeneration of 3 micronuclei in each conjugant → Mitotic division produces 2 micronuclei each from the remaining one micronucleus, i.e. one immotile 'female' micronucleus and one migrant 'male' micronucleus → Following exchanging migration, fusion occurs between the immotile and migrant micronuclei, as in *Paramecium aurelia* → Separation of conjugants → Formation of 2n zygote by fusion between 2 micronuclei in each of the ex- conjugant → Production of eight micronuclei by three consecutive mitotic divisions → four of them become micronuclei and the other four macronuclei → Formation of two progenies each with one micro- and one macro-nuclei, i.e. out of two conjugants, only two progenies are produced or two progeny/ conjugant (Fig. 1.23).

In ciliates, the clonal multiplication cycles are more often interrupted by endomixis, a process involving nuclear reorganization. The interval between two successive endomixes ranges from 25 to 35 days (but 4 to 129 days in *Paramecium aurelia*) and is accompanied by a decline in fission rate and increased volume, opacity, viscidity and sluggishness of the individual – a state called depression. In it, the macronucleus fragments and disappears. The micronucleus divides many times followed by disappearance of most of its products and the remaining ones reconstitute new micro- and macro-nuclei that are distributed by means of fission until normalcy is regained. In many ways, the events in endomixis resemble those in conjugation, except for the fusion of haploid micronuclei. The incidence of endomixis within a cyst has been studied in *Dileptus, Didinium, Spathidium, Stylonychia* and *Uroleptus*. In a way, endomixis may rejuvenate the clones but its absence in many ciliates fails to confirm rejuvenation, albeit the micronuclear fission and associated events may generate new gene combinations.

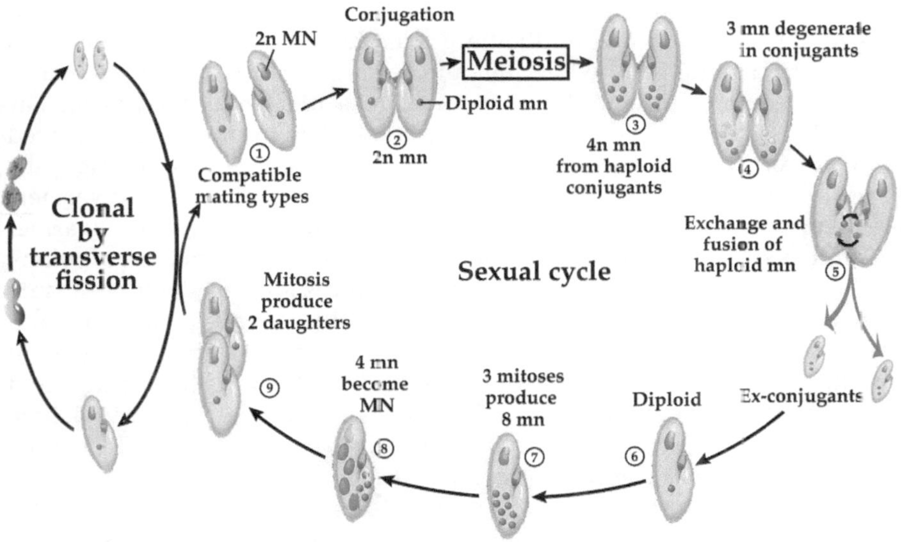

FIGURE 1.23

Life cycles in a typical ciliate. Clonal and sexual reproduction by conjugation in *Paramecium* (Anderson. 1988). MN = macronucleus, mn = micronucleus. Note all the major stages are marked by rounded numerals.

On the other hand, conjugation is a necessary rejuvenating process. Notably, binary fission by itself introduces some rejuvenation and reduces senescence to a certain extent. Thus, it was possible for Woodruff (1926) to maintain a clone of *P. aurelia* for over 1,500 generations without decline in vitality or fission rate, although the clonal cycle was interrupted by endomixis at least once after every 303 generations. Fenchel and Finlay (2006) reported that *Cyclidium glaucoma* can be maintained clonally for several thousand generations. A few ingeniously designed experiments by early authors have clearly shown the obligate need for the interruption of long successive clonal cycles by conjugation. Thus, a study involving clones from exconjugants and non-conjugants – both derived from the same original clone of *Uroleptus* and *Spathium* revealed that (i) conjugants indeed prolonged the lifespan of a clone, i.e. postponed extinction of a clonal lineage and (ii) accelerated fission rate, i.e. prevented the onset of senescence. Another study involved the separation of conjugants prior to micronuclear exchange and found the rejuvenation effect of conjugation. On the whole, conjugation is an obligately required process to interrupt the unusually long repeated clonal cycles of ciliates. *Both conjugation and endomixis provide the scope for the generation of more new gene combinations than in the sexual cycle of mastigophores and rhizopods*. These have contributed a great deal to the species diversity in Ciliophora, the second most speciose groups among protozoans.

In a vast majority of euciliates, the conjugants are isogamous, i.e. similar in size and shape. However, they are anisogamous, i.e. one is smaller in size the microconjugant, while the other is larger called macroconjugants. They are common among the parasitic oligotrichid family Ophrycolicidae (Fig. 1.24C) and in sessile peritrichs (e.g. *Vorticella*). In the former, represented by *Opisthotricium*, the conjugation cycle, however, passes through the same one to nine stages. In the sessile vorticellids, the motile swarmers are generated clonally to facilitate dispersal; on settling, the swarmer undergoes successive vertical divisions and leads to the formation of colony (Fig. 1.24A). In the other, a bud from the 'bell' turns into a swarmer to ensure dispersal. In them, conjugation is modified by the generation of micro- and macro-gamonts from different populations. The entry by the microgamont into the macrogamont (Fig. 1.24B) is followed by the nuclear exchange and the usual meiosis and fertilization. Despite one or another minor variation, *sexual reproduction by conjugation is perhaps the only life cycle type in all euciliates, that clearly indicates monophyletic nature of ciliates*.

A rare but interesting life cycle of the 2-hosted parasitic ciliate is also described. Typically, the cycle of the vermiform *Chromidina* involves a definitive host (DH), the euphausiid *Nematoscelis* and an intermediate host (IH) the squid *Pterygioteuthis*. In the latter, the monotomic binary fission or preferably budding of the posterior body results in clonal multiplication. In the IH euphausiid, multiple budding is followed by conjugation (Fig. 1.25).

Apart from these, the life cycle of the Opalinata – as represented by *Opalina ranarum* is also included in order to complete the description on life cycles of ciliates. As in euciliates, the opalinids are also ciliated but they lack nuclear duality and conjugation. For example, *O. ranarum* is a non-pathogenic parasite in the gut of frogs and tadpoles. The clonal cycle takes place in the adult gut. However, its sexual cycle occurs in tadpole 2, as illustrated in Fig. 1.26. Notably, species number of the Opalinata is very few (< 200, *Wikipedia*) that they may have to be relegated to a very small proportion of ciliates.

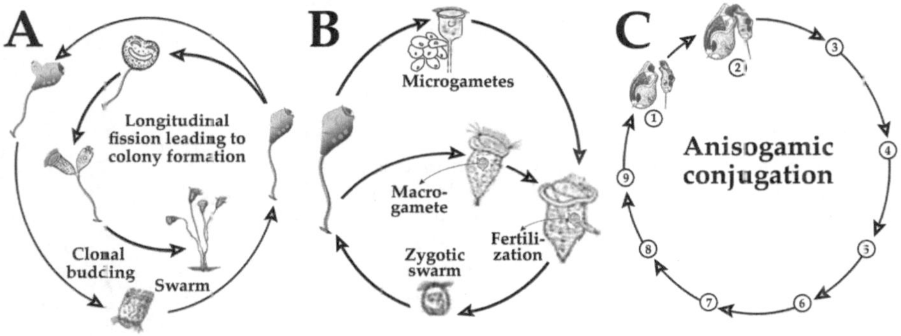

FIGURE 1.24

(A) Clonal budding and longitudinal fissions lead to colony formation in *Vorticella* (B) Sexual reproduction in *Vorticella*. (C) Presumed anisogamic conjugation in a parasitic ciliate *Opisthotricium* (based on Hyman, 1940). In Figure C, the rounded numerals indicate the stage, as in Fig. 1.22.

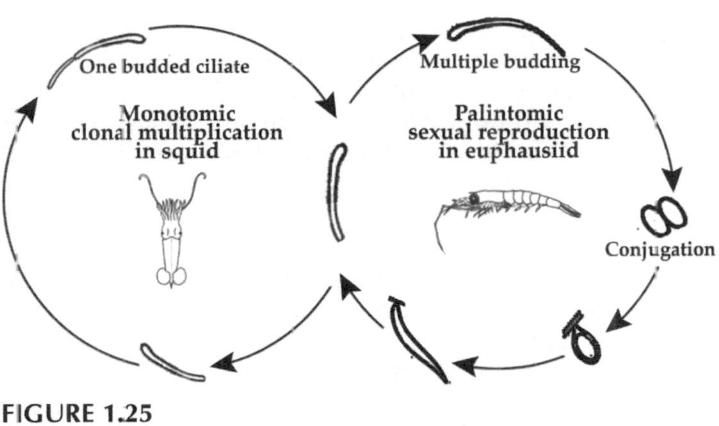

FIGURE 1.25

Life cycle of *Chromidina cortezi* involving palintomic sexual reproduction in eupahsuiid *Nematoscelis difficilis* and monotomic clonal multiplication in squid *Pterygioteuthis giardi* (modified and redrawn from Hochberg, 1983).

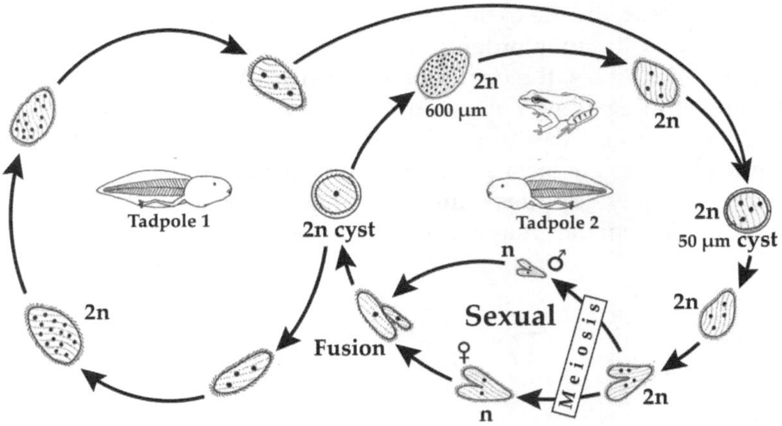

FIGURE 1.26

Life cycle of *Opalina ranarum*. Ingestion of defecated 2n cyst by tadpole 2 → sexual life cycle results in production of 2n cyst, which may be ingested by adult frog. Alternatively, the cyst may enter tadpole 2 and ultimately passed as 2n cyst (modified and redrawn from Mehlhorn, 2008).

1.4 Motiles – Sessiles – Colonials

One reason for colony formation is to increase the size and thereby escape predation. Computer searches have yielded the required values for the colonials and sessile mastigophores and rhizopods. However, it has brought only incomplete information for ciliates. The survey involving many sources (Hyman, 1940, taxonomic description by Levine et al., 1980, pictorial representations of protozoan taxonomic groups by Warren et al., 2016 and computer searches) also yielded only some data. However, it must be noted that (1) the values drawn from WoRMS are underestimates, the cumulative total species number listed in WoRMS is only 3,165 for protozoan, while their actual number, as of today is 32,950 (Table 1.15). (2) Values were collected only for those genera that are pictorially represented by Hyman (1940) and Warren et al. (2016). There may be more genera consisting of sessile protozoans. With these constraints, the approximate values are assembled in Table 1.12. Hence, the values reached at are certainly underestimates. In his reassessment of number of species/genus and so on, Corliss (1977) reported that of 1,234 ciliate genera, > 50% of them are characterized by one species/genus. On the whole, these values may not exceed ~ 1230 and 920 species for the colonials and sessiles, respectively.

The survey revealed that (i) among radiolarians, the group Spumellaria consist of 244 species (*eol.org*) and may include the species belonging to the genera *Sphaerozoum* and *Colosphaera*. Being colonies, they engage zoochlorella as symbionts (Suzuki and Not, 2015), (ii) colonial occurrence is also known from two orders of Mastigophora namely (a) the Volvocida and (b) Choanoflagellida, as well as from (c) Suctoria and (d) Peritrichida among Ciliophora. Most surprisingly, the survey also revealed for the first time that all mastigophoran volvocid and rhizopodan radiolarian colonies are motile floatants, whereas those of choanoflagellates and ciliates are sessiles – a finding known from the days of Hyman (1940) until those of Warren et al. (2016). But none has ever recognized it.

Among Mastigophora (i) the Volvovida group consists of motile 341 solitary species and 52 colonial species; they are freshwater inhabitants. The motile floating gelatinous volvocid colonies hold 4–16 zooids, as in *Gonium*, oval with 32 zooids, as in *Platydorina* or spherical consisting of 128 zooids, as in *Pleodorina* (Hyman, 1940). Apparently, the spherical shape facilitates holding increased number of zooids in a colony. Except in *Pleodorina*, each constitutional zooid in a colony is capable of clonal or sexual reproduction. These colonies are polarized, as they swim always through one or anterior pole. But in the exceptional *Pleodorina*, the zooids at the anterior pole are sterile and differ morphologically from the posterior reproductive zooids, indicating that these colonies approach metazoan status. In them, clonal multiplication may involve division of mucilage sheath rather than longitudinal fission of zooids in specific rows.

Among *Volvox*, each colony has a species-specific number of zooids, constant size (up to 2 mm) and shape. The number of which ranges from 500 to 50,000. The zooids are arranged on hollow gelatinous sphere and each of them has its own mucilage sheath. Daughter colonies arise clonally from buds or gonadia. Each gonadium undergoes 11 successive longitudinal fissions to produce a daughter colony. In the colonies, the 32-kDa glycoproteinaceous pheromone acts as chemoattractant of gametes even at concentration below 10^{-16} M (see El-Bawab, 2020).

In choanoflagellates, the stalked sessiles may be solitary or colonial. Of 125 species, surprisingly, 108 are solitaries and only 17 are colonials. The colonials like *Proterospongia* (Fig. 1.3K, 1.27B) consists of a gelatinous mass of collared zooids at the periphery, but the central core of collarless zooid called amoeboid zooids that are derived from surface zooids wandering inwardly. As the name suggests, *Proterospongia* and its relatives are regarded as a link between choanoflagellates and sponges (Hyman, 1940). *The rare combination of coloniality and motility cum floating is limited to 52 volvocid species or 1% of all*

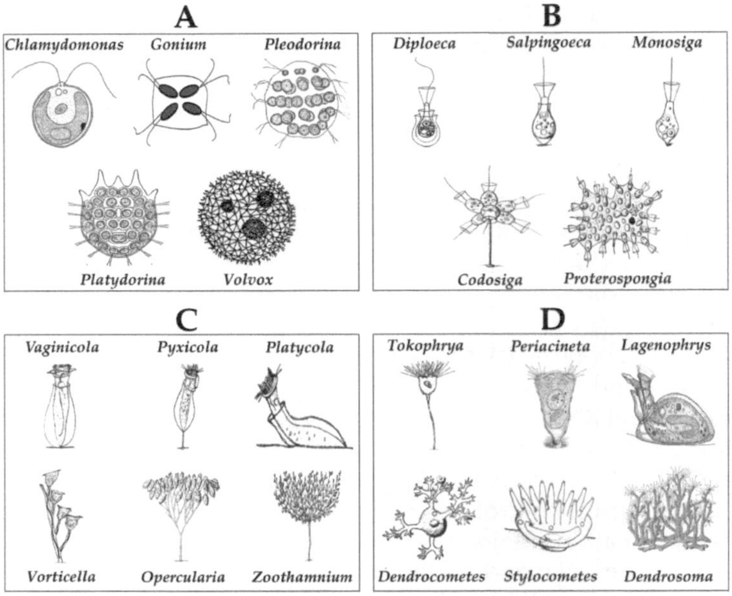

FIGURE 1.27

Representative species of (A) solitary and colonial motile volvocids, (B) solitary and colonial sessile in choanoflagellates, solitary and sessiles (C) peritrichidan and (D) suctorian ciliates (adopted from early authors, as listed in Hyman, 1940, Warren et al., 2016).

mastigophores, 244 species or 2% of all rhizopods. Understandably, these autotrophs or radiolarians harboring symbiotic zoochlorella have to float and keep moving to capture adequate sun light. Sessility may reduce the cost of motility but may expose them to predation. Further, it limits the volume of habitat water, from which food has to be acquired. *Not surprisingly, only 125 choanoflagellate species, i.e. 1.2% mastigophores and 916 colonial ciliates species, i.e. 11.5% of ciliophores have opted for sessility.* With 3-times increased incidence, *the ciliature and mode of life among peritrichids* (Fig. 1.27C) *and suctorians* (Fig. 1.27D) *have rendered ciliates to be better adopted for sessility than choanoflagellates* with a collar holding a flagellum. Interestingly, > 200 vorticellid colonial species are found as either commensals or ectoparasites on the motile crustacean body (Mayen-Estrada and Utz, 2018). These vorticellids are ingenious, as they are sessiles but motiles at the cost of their motile hosts.

By virtue of their size, colonials may avoid predation but it increases intraspecific competition among colonial constituents for food and oxygen. Not surprisingly, coloniality is limited to (244 radiolarians + 52 volvocids + 17 chaonoflagellates) 69 species or 1.2% of all flagellates, and 2.2% of all rhizopods, in comparison to (683 peritrichids + 233 suctorians) 916 colonial species or 11.5% of all ciliates (Table 1.12). Hence, *ciliates are also better adapted for* *coloniality than flagellates. To avoid intraspecific competition, 86% Mastigophora (i.e. 449 of 518 species) have adopted solitary sessility, whereas only < 5.6% (452 of 8,000 species) ciliates have opted for it.* Within ciliates, the peritrichids are better adapted to sessility either as solitaries (338 species) or as colonials (683 species) than their counterpart suctorians (114 solitaries, 233 colonial species). On the whole, *the colonials (52 volvocids + 17 choanoflagellates+ 244 radiolarians + 916 ciliates) constitute 1229 species or 3.7% of protozoans. With 1,494 species, sessility*

TABLE 1.12

Estimation of the number of sessile, solitary and colonial protozoans. Values are drawn from *onezoom.org*. * Pandian (2022). † Underestimated values (see p 109)

Taxa	Species (no.)		
	Solitary	Colonial	Total
Volvocida	341	52	393
Choanoflagellates	108	17	125
Subtotal	449 (86.6%)	69 (13.4%)	518
Radiolarians	–	244	244
Ciliates			
Peritrichids	338	683	1021
Suctorians	114	233	347
Subtotal	452 (33.0%)	916 (67.0%)	1368
Grand total	901 (42.3%)	1229 (37.7%)	2130
Metazoans*	1,490 (04.8%)	29,647 (95.2%)	31,137

Peritrichia†				Suctoria†			
Solitary		**Colonial**		**Solitary**		**Solitary**	
Cothurnia	163	*Carchesium*	7	*Acinetopsis*	1	*Podophrya*	37
Lagenophrys	79	*Campanella*	40	*Choanophrya*	1	*Prodiscophrya*	1
Lophophorina	1	*Vorticella*	287	*Discophrya*	2	*Pseudogemma*	3
Platycola	4	*Ophrydium*	23	*Echinophrya*	1	*Solenophrya*	13
Pyxicola	1	*Epistylis*	168	*Ephelota*	8	*Spelaeophrya*	1
Rhabdostyla	50	*Opercularia*	88	*Heliophrya*	3	*Squalorophrya*	2
Stentor	22	*Zoothamnium*	70	*Lecanophryella*	3	*Stylocometes*	1
Thuricola	14	Subtotal	683	*Metacineta*	17	*Tachyblaston*	1
Vaginicola	4	Total for Peritrichia	1021	*Multifasciculatum*	5	*Thecacineta*	4
Subtotal	338			*Periacineta*	7	*Tokophrya*	1
				Podocyathus	2	Subtotal	114

Suctoria†					
Colonial	**(no.)**	**Colonial**	**(no.)**	Total for Suctoria: 114 + 233 = 347	
Acineta	127	*Metacineta*	8	**Total for Ciliates: 1021 + 347 = 1368**	
Cometodendrum	1	*Paracineta*	33	Mastigophores	518
Dendrocometes	4	*Scyphidia*	59	Radiolarians	244
Dendrosoma	1	Subtotal	233	**Grand total**	2130

constitutes 4.5% of protozoans. Interestingly, colonies need not necessarily be sessiles; there can be motile colonies. Similarly, sessiles need not necessarily be colonials; there can be solitary sessiles.

1.5 Encystation – Excystation

As in other organisms, encystation is a strategy adopted by both free-living and parasitic protozoans to escape unfavorable circumstances like lack of food and/or oxygen, drying or freezing or a fouling medium due to accumulation of metabolic wastes. Two types of encystations are recognized: Voracious feeding is followed by digestive encystation, apparently to accommodate the slow digestive process. The second one is called reproductive encystation, in which sexual (gamogonic) or clonal (sporogonic) multiplication is accomplished (Hyman, 1940). The ensuing description elaborates only the reproductive encystation. In free-living protozoans, encystation facilitates (i) genome transfer from one generation to the next and (ii) dispersal from one habitat to another. It also serves for the transmission from one host to another, besides the genome transfer through successive generations in parasitic protozoans. The encysted protozoan stage is therefore a dynamic one, unlike in others, in which, the cyst may represent a quiescent stage. In protozoans, the unfavorable circumstance-induced entry into the dynamic cyst stage is followed by a short (e.g. 6 hours in the tidal oligotrich ciliate *Strombidium oculatum*, see Verni and Rosati, 2011) or a longer duration lasting for a few years (e.g. 49 years, *Cercomonas*, see Hyman, 1940). Following excystation, the emergence and release of a few or more protozoan offspring occur under favorable conditions.

Encystation involves production cost but provides safer transfer/ transmission. Almost all the 1,650 speciose Gregarinia and single hosted 1,445 speciose eucoccidian Eimeriina release only encysted sporonts (Table 1.9); in the encysted sporonts, sporogony takes place. Production of naked sporozoites seems to be a characteristic feature of the two hosted eucoccidian haemogregarines and haemosporinans. Both of them are corpuscular parasites. Surprisingly, the single hosted Haplosporea also produce naked gregarious sporozoites from a cyst (Fig. 1.20), which are transmitted from one host to other in filter-feeding oyster. The filter-feeding hosts like bivalves, mosquito larva and probably copepods seem to acquire naked sporozoites through aquatic medium. In the two-hosted Microsporidea, represented by *Amblyospora* and Coccidia (e.g. *Aggregata eberthi*), gamogony generates naked gametes. On the whole, 60, 9 and 31% sporozoans produce encysted sporonts, naked sporozoites and both of them, respectively (see Table 1.10). Production cost seems to be a constraint in the two-hosted taxonomic groups, as they have opted for naked ones in one of the hosts.

TABLE 1.13

Estimation on the number of encysted protozoans (species numbers are drawn from *ucl.ac.uk, onezoom.org*). ‡ see text

Taxa	Species (no.)	Species (%)	Taxa	Species (no.)	Species (%)
1. Encystation after fusion/fertilization					
(i) Mastigophora			(ii) Rhizopoda		
a. Chryptomonadids	50		a. Amoebida	181	
b. Chrysomonadids	50		b. Arcellinida	2000	
c. Coccolithophora	50		c. Heliozoa	100	
d. Chloromonadida	500		d. Parasitic rhizopods	250	
e. Volvocida	400		Subtotal	2531	
f. Dinoflagellida	2000		(iii) Sporozoa		
g. Euglenida	1000		a. Haplosporea	51	
h. Choanoflagellida*	125		b. Eucoccidia	1445	
Subtotal	4175		Subtotal	1496	
			Total	**8202**	**24.9**
2. Encystation prior to fertilization					
Radiolaria	4200		**Total**	**4400**	**13.4**
Opalinata	200				
3. Encystation only at or after clonal multiplication					
(i) Mastigophora			(iii) Sporozoa		
a. Cryptomonadida Chrysomonadida Coccolithophorida	150		a. Gregarinia b. Coccidia c. Microsporidea	1650 712 1300	
b. Parasitic flagellates	1800		Subtotal	3662	
Subtotal	1950		(iv) Ciliophora	7800	
(ii) Rhizopoda					
a. Amoebae	19		**Total**	**14374**	**43.6**
b. Foraminifera‡	943				
Subtotal	962		**Total for encystation**	**26976**	**81.9**
4. Non-encysted protozoans					
Foraminifera	3557		Haemosporina	156	
Trypanosoma spp	100		**Total**	**3898**	**11.8**
Haemogregarina	85		**Grand total**	**30874**	**93.7**

Based on Figs. 1.7 to 1.26, the timing of encystation can be considered under the following four groups: 1. At or after fusion/fertilization, 2. Pre-fertilization, 3. At or after clonal multiplication and 4. No encystation. Heinz et al. (2005) brought evidence for the occurrence of agglutinate microspherical (e.g. *Ammonia beccarii, Rosalina bradyi*) and calcareous (porcelain, e.g. *Adercotryma catinus*) cysts in benthic foraminifers. Fortunately, the species number for the agglutinated and calcareous benthic foraminifers is reported by Murray (2007). For choanoflagellates, the incidence for cysts in freshwater and marine habitats is hinted by Leadbeater and Karpov (2000) and Karpov (2016). In his lengthy review, Noble (1944) is silent on encystation in Microsporidea. However, their incidences are indicated in freshwater (e.g. *Henneguya*, Hyman, 1940) and marine (e.g. *Ceratomyxa*) fishes (Lom, 1984). In all, relevant information for encystation has been assembled for 30,874 species or 93.7% protozoans. Approximately, *3,900 or 11.8% protozoans do not encyst* (Table 1.13). In 256 species, transmission is achieved by sanguivorous leeches or insects. More than 3,500 foraminifers do not encyst, as their wavy pelagic habitat provides them adequate food and ensures fertilization but leaves 943 benthics to encyst. In the remaining *26,976 species or 81.9% protozoans encyst. Encystation occurs in Group 1 after fusion/fertilization in 8,202 species (24.9%), Group 2, prior to fertilization in 4,400 species (13.4%) and Group 3, 14,374 species (43.6%) at or after clonal multiplication.* However, it must be noted that taxonomic groups within Group 1 may also undertake successive clonal multiplications. In them, sexual reproduction may be a rare event and when it occurs, it produces only a few offspring. In *Amoeba*, for examples, the subitaneous binary fissions results in two daughter progenies but multiple fissions within a cyst generate as many as 32 offspring. Interestingly, within the parasitic flagellate genus *Giardia*, *G. vaginalis* do no encyst but *G. intestinalis* do encyst and produce as many as 10^{10} cyst/d, although the infective dose is as low as 25–100 cysts (Dalton et al., 2001). *The encystation is an early discovery of the protozoans and has the longest history and has repeatedly appeared many times, for example, 1 BYA in radiolarians and < 5 MYA in parasitic flagellates like Giardia in human.*

That most protozoans encyst indicates the survival benefit arising from encystation. For example, an experimental culture has shown that of 3,000 individuals of the ciliate *Oxytricha bifaria*, an average of 33% died during vegetative phase, in comparison 12% mortality during encysted stage. Immediately following excystation, another 3% succumbed to death. Hence, the cumulative mortality during and immediately following encysted stage may not exceed 15%, i.e. mortality suffered due to a vegetative stage is nearly two-times higher than that during encystation. Not surprisingly, *82% protozoans have opted the strategy of including encystation in their life cycle to ensure genome transfer through generations and dispersal from one habitat to other in free-living species and/or transmission from one host to other in parasitic species.*

The cyst formation includes two successive steps: (1a) cytoplasmic reorganization, (b) along with reduction in volume and (c) transformation to spherical shape and (2) formation of the cyst wall. The cytoplasmic reorganization listed for ciliates by Verni and Rosati (2011) may also hold true for other protozoans. Accordingly, the reorganization comprises the following: (a) storage and localization of nutrients to sustain the dynamic cyst, (b) resorption of locomotory organelles, (c) cytoplasmic condensation and ribosomic enrichment, (d) grouping of dispersed mitochondria into cluster(s) and (e) macronuclear fusion and its arrangement into a small spherical dense body in ciliates. (3a) In the reduction of the cell volume, the contractile vacuole plays a key role. The frequency of vacuolar discharge ranges from 30 to 120 times; it is dependent on the level, to which the reduction is accomplished and the stage, from which encystation proceeds. Incidentally, *the vacuolar presence is generally considered as a characteristic of freshwater protozoans (Hyman, 1940). However, the genes responsible for the formation and functioning of the vacuole must also be present in marine species but they are expressed only during encystation*. The levels of volume reduction (leading into cytoplasmic condensation) ranges from 3% of the initial vegetative size in the heterotrich *Condylostomides etoschensis* to 97% in the haptorid *Arcuospathidium cultriforme* (Table 1.14). However, no reduction occurs in another heterotrich *Blepharisma japonicum*. In fact, in some like *Strombidium oculatum*, the volume is increased by 1.5 times of its vegetative size. (3b) The volume reduction is accompanied by transformation into spherical shape, which reduces the surface area to the minimum, over which the cyst wall is to be formed. For example, surface area of the elliptical *Oxytricha* is reduced by five times or 20% of the imminent vegetative shape. Ricci et al. (1985) had calculated that even if the cyst has the same volume like the vegetative one, as

TABLE 1.14

Reduction in cyst volume as percentage of vegetative animal volume in some Ciliophora (from Verni and Rosati, 2011)

Species	Cyst volume (%)	Species	Cyst volume (%)
Oligotrichs		**Peritrichs**	
Meseres corlissi	28	*Opisthonecta henneguyi*	24
Pelagostrombidium spp	58	*Vorticella echini*	51
Strombidium oculatum	161	**Colpodids**	
Stichotrichs		*Maryna umbrellata*	52
Oxytricha bifaria	20	*Colpoda cucullus*	69
Parakahliella halophila	26	*Kuechneltiella namibiensis*	84
Engelmaniella mobilis	33	**Haptorids**	
Hetrotrichs		*Enchelydium blattereri*	66
Condylostomides etoschensis	3	*Arcuospathidium cultriforme*	97
Blepharisma americanum	92	*Spathidium turgitorum*	100
B. japonicum	100		

in *B. japonicum* or the haptorid *Spathidium turgitorum*, the spherical shape ensures a saving of ~ 75% surface area to be covered by the cyst wall. (4). An ultrastructural study in the ciliate *Oxytricha bifaria* has shown that its cyst wall consists of four distinct layers (i) the ectocyst, (ii) ornamented (with finger-like protrusions) external lamellar layer, (iii) homogenous internal lamellar layer of similar shape and thickness and (iv) granular endocyst. A similar four layered configuration of the cyst wall has been reported from many ciliates (e.g. *Gastrostyla steini*, Rosati et al., 1983). The number of frequency of finger-like protrusions is 80–100/cyst, which renders the external surface irregular and thereby avoid predation.

Publications on protozoan cyst are widely scattered, and typically limited to their incidence (Corliss and Esser, 1974) but scarce description of their morphology (Fig. 1.28). Figure 1.28J to N represents the developmental sequence in cyst of *Cryptocaryon irritans*. Some like *Woodruffia metabolica* have two cyst types – a stable one resistant to drying and the other unstable, which cannot tolerate drying. In others, the exposure to 50°C kills the wet cyst but the dry ones can tolerate up to 120°C. The cyst of *Colpoda maupasi* can withstand twice the amount of ultra-violet irradiation than the live ones. Starvation is demonstrated as the most important inducer of encystation. In the fields, low pH of 6.4 induces encystation but that of 8.4 excystation (Corliss and Esser, 1974). The excystation requires just 10 minutes for *Giardia lamblia* and *G. muris* and is facilitated at pH 4.0 (Adam, 2001). The melatonin level increases by several orders of magnitude during encystment and may prevent oxidation of the lipids in the cyst. Calcification of the cyst in some genera occurs by the deposition of calcium carbonate crystals in the narrow space between the cell wall and plasma membrane. *Ceratium hirudinella* contains silicon layers. The cyst wall of the dinoflagellate is extremely resistant to decays, as it contains dinosporin, a substance similar to sporofellinin in the pollen of higher plants.

1.6 Shells and Skeletons

The locomotion speed ranges from 2 to 3 µm/s for the rhizopods, from 15 to 300 µm/s for the flagellates and from 400 to 2,000 µm/s for the ciliates (Hyman, 1940). To escape from predation, the need for a shell and skeleton is a critical requirement more for rhizopods than for ciliates. Within the rhizopods, the 4,500 speciose foraminifers, 2,000 speciose arcellinids and 300 speciose filosians are all shelled, i.e. 6,800 speciose or 59% rhizopods are testated or

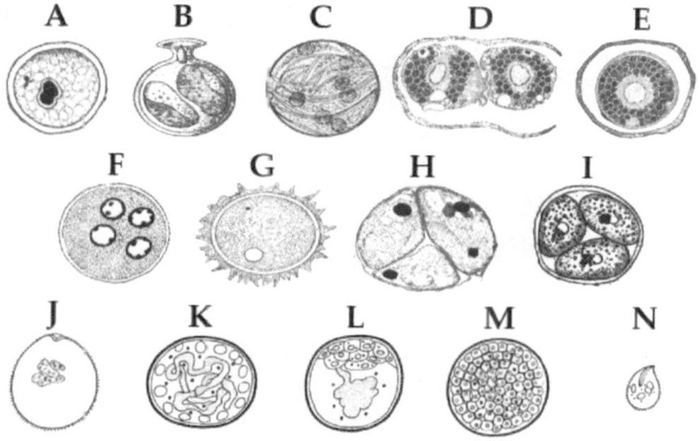

FIGURE 1.28

Representative cysts. (A–C) Mastigophores: A = hologamic encysted zygote of *Copromonas subtilis* (after Dobell, 1908), B = pot-like cyst of *Chromulina* (after Conrad, 1926), C = crescentic cyst of *Gymnodinium lunula* (after Kofoid and Swezy, 1921); (D–F) Rhizopods: D = hologamic cysts showing formation of vegetative gamete within the cyst of *Actinophrys*, E = the same cyst after completion of hologamy (after Belar, 1923), F = four-nucleated cyst of *Entamoeba histolytica* (after Dr. D.H. Wenrick); (G–I) Ciliophora: G = *Pleurotricha* (after Ilowaisky, 1924), H = fission within the cyst of *Tillina* (after Gregory, 1909), I = cyst undergoing sexual multiplication in *Ichthyophthirius multifilis* (after Rhumbler, 1888). J–N represent the development within the cyst of *Cryptocaryon irritans*. Note the ribbon-like macronucleus and numerous small dark micronuclei in K, unequal division in L, cyst full of small domites in M and released infective theront in N (redrawn after Brown, 1963).

protected by the calcareous shell. As in gastropods, the foraminiferan shells are multi-chambered and the animal inhabits the outer most chamber (Fig. 1.4 G). As mentioned earlier, Mikhalevich (2021) describes ~ 130 different shell types.

The testate rhizopodeans are characterized by a single chambered inverted shell (Fig. 1.4C–E). Among the actinopodean rhizopods, the skeleton of radiolarians is composed of hydrated, opal, oxidized silicon compounds. On it, a thin layer of divalent cations such as Ca^{2+} may be incorporated (Suzuki and Not, 2015). Added with the 4,200 speciose skeletonous radiolarians (Fig. 1.4K, L), ~ 95% of rhizopods are protected by a shell or skeleton. The thecate dinoflagellates are also encased by cellulose plates (Fig. 1.3E), which are often further strengthened by various ornamentation.

In general, most free-living protozoans are endowed with a capacity to secrete tectin, a mucus-like substance. Some are enclosed by more or less permanent coverings, which may consist of organic material; the covering may further be strengthened by minerals, soils, silica or calcium carbonate. Others like the flagellates, may be enveloped by a coat of gelatinous or a tectinous material. However, these coverings/envelops may readily be lost and reformed.

1.7 Species – Number – Diversity

Species: The concept of species and speciation leading to biological diversity is of great importance to the theory of evolution. Relevant information on species diversity of protozoans is scattered throughout the foregone description and it shall not be described again. According to Ernst Mayr (1942), a species is a group of actually or potentially interbreeding population that is reproductively isolated from all other groups (see Pandian, 2021b). This definition, though acceptable to zoologists, is not readily acceptable to botanists. For (i) some 23% of polyploid angiosperms obligately undergo clonal multiplication (ii) plants are more amenable to hybridization than animals and (iii) uniquely plants are also amenable to interspecific grafting as well as somatic embryogenesis (see Pandian, 2022). Protozoologists are also hesitant to accept Mayr's definition, as (i) some protozoans like free-living polyploid amoebidae and diploid trypanosomids reproduce only through clonal multiplication and so are euglenids as well as parasites and symbiotic flagellids that belong to five orders of Zoomastigophora, (ii) the testate rhizopodeans like *Difflugia* inherit differently the shelled and shell-less traits through clonal multiplication (Fig. 1.9A, see also Poljansky, 1992). It is in this context, the definition by T.M. Sonneborn is more acceptable to protozoologists. Accordingly, a species is an evolving entity that has undergone a threshold minimum evolutionary divergence (see also Schlegel and Meisterfeld, 2003). From the studies on many isolates of heterotrophic dinoflagellate *Oxyrrhis marina* collected from different oceans and seasons, Lowe et al. (2005) found no evidence for a geographical pattern, implying that the clonal lineages may not lead into speciation. Fenchel and Finlay (2006) also considered the protozoan species as the basic taxonomic unit within the taxonomic hierarchy.

TABLE 1.15

Approximate estimate for species number among major groups of Protozoa (Levine et al., 1980, Adlard and O'Donoghue, 1998†, *parasite.org.au**, Lynn, 2010a‡)

Group	Existing	Extinct
Mastigophora	6,900†	Not known
Rhizopoda	11,550	47,500†
Sporozoa	6,500*	–
Ciliophora	8,000‡	200*
Total	32,950	47,700

7800 radiolarian species + 40500 foraminiferan species (Corliss, 2001)†

Number: More than any organismic groups, the protozoan taxonomy remains fluid, tardy and inconsistent. The following are some reasons for these: (i) Lack of sufficient morphic features, (ii) difficulties encountered in culturing them to describe the life cycle, (iii) lack of methods to prepare and store reference material in musea and (iv) inadequate number of trained taxonomists (Warren et al., 2016). Since the inclusion of two protozoan species in *Systema Naturae* by Carl Linnaeus in 1674, their number has increased to 3,165 (WoRMS), 31,220 (Adlard and O'Donoghue, 1998), to 32,000 (*parasite.org.au*) and to 200,000 (Cox, 2002). As most reported values are around 32,000 species, the value arrived is 32,950 species for Protozoa. However, the estimate of 47,700 extinct species may (Table 1.15) not still be acceptable, as relevant information is not available for the two major groups other than Rhizopoda and Ciliata.

2

Spatial Distribution

Introduction

The oceans cover 70% of the earth's surface with 97% of its water. Freshwater systems, however, cover only ~ 1% and hold as little as 0.01% of its water. The remaining 29% of the earth's surface is covered by land (see Pandian, 2011). With water masses of 1.36 billion km^3 (*jbuttler@uh.edu*), the oceans provide 900-times more livable volume of space than in land. Temperature and precipitation are decisively important factors that provide the scope for speciation and diversity in land. Average annual precipitation over the earth is $5.77 \times 1014\ m^3$, of which 79% fall over the oceans and only 21% on the land (Agrawal, 2013). Receiving less precipitation than evaporation of water, the deserts with over 28.5 million km^2 (*edu.seattleepi.com*) span over 9% of the land and leave only 20% land area with more precipitation than evaporation. Hence, productivity and biodiversity occur in the remaining 20% of the land alone.

In aquatic systems, light penetration diminishes with increasing depths. In the absolute absence of any light below 2,400 m depth, no photosynthesis occurs. Consequently, absence of light effectively reduces the availability of nutrients and dissolved oxygen. However, a large fraction of dead and decaying organisms sinks to the bottom, which forms the major nutrient source for metazoans and protozoans. Incidentally, the latter depend mostly on bacteria that decompose the sinking organisms. Nevertheless, only a few radiolarian species are reported from the depths of ~ 8,000 m, although they are more abundant and diverse between 200 and 2,000 m depths (Anderson, 1983). Feeding on bacteria and detritus, the agglutinated komokaicean foraminifers with complex anastomosing network of tubules are known from up to ~ 4,000 m depth (see Finlay and Esteban, 2018). Except for these two rhizopod groups, others, especially the mastigophores and ciliates are not known from the abyssal.

2.1 Horizontal Distribution

Protozoa inhabit water, wherever it is available. Though they are predominantly aquatic inhabitants, some are also found in moist soils, on damp mosses and others. They are reported from –4°C in icy habitats to 40–65°C in hot springs. Some protozoans are known to survive up to 70°C on slow acclimation to increasing temperature over a period of > 7 years (see Hyman, 1940). Even among the exclusively marine 4,500 speciose Foraminfera, there are freshwater inhabitants like *Gromia, Allogromia* and *Lieberkuhnia*. In *Thorp and Covich's Freshwater Invertebrates*, Damborenea (2020), Kosakyan et al. (2020) and Kuppers et al. (2020) do not even hint at the species number of freshwater flagellates, rhizopods and ciliates, respectively. Repeated computer searches have failed to yield adequate data on spatial distribution of protozoans. However, Finlay and Esteban (2018) estimated the approximate number of protozoan species inhabiting the marine and non-marine habitats. However, their estimate is limited by the following: (i) It does not include the 6,500 speciose Sporozoa. (ii) Even within the other three major groups, it is limited to 11,130 (out of 26,450) species or 42% only. (iii) It requires separate estimates for freshwater and terrestrial species, although the latter may contribute a few species alone. According to their estimate (Table 2.1),

TABLE 2.1

Estimates on marine, freshwater and terrestrial protozoans. Estimate by Finlay and Esteban (2018)†, Foissner (2014)* are included. Values for Sporozoa from Table 1.10

Class	Marine		Non-marine		†	Total	
	(no.)	(%)	(no.)	(%)		(no.)	† as (%)
Mastigophora	1650	78.1	460	21.8	2110	6900	30.6
Rhizopoda	5420	90.9	540	9.1	5960	11550	52.0
Ciliophora	1400	45.8	1660	54.2	3060	8000	38.0
Subtotal	8470	76.1	2660	23.9	11130	26450	42.1
Estimated values for 26,450 protozoan species on applying percentage values of Finlay and Esteban (2018)							
Class	Marine		Freshwater		Terrestrial*		Total
	(%)	(no.)	(%)	(no.)	(%)	(no.)	
Mastigophora	78.1	5390	16.7	1150	5.2	360	6900
Rhizopoda	90.9	10500	6.8	780	2.3	270	11550
Ciliophora	45.8	3660	41.8	3340	12.5	1000	8000
Subtotal	73.9	19550	19.9	5270	6.2	1630	26450
Estimated values for 32,950 free-living and parasitic protozoan species (Table 1.10)							
Sporozoa	57	3710	11	718	32	2071	6500
Total	70.6	23260	18.2	5988	11.2	3701	32950

approximately 78% (or 1,650 species), 91% (or 5,420 species) and 46% (or 1,400 species) Mastigophora, Rhizopoda and Ciliophora are marine inhabitants, respectively.

To get a complete picture on spatial distribution for the three classes of protozoans, the following two steps were adopted. First , the values reported by Finlay and Esteban (2018) for 11,130 species were converted to 26,450 species by applying their percentage values. Second, the species number for the freshwater and terrestrial protozoans were separated. Fortunately, Foissner (2014) reported the class-wise species number for the terrestrial protozoans (Table 2.1). On subtraction from the converted values by the reported values for the terrestrial protozoans of the three classes, the values for the freshwater protozoans were obtained. Accordingly, 1,150 species (16.7%), 780 species (or 6.8%) and 3,340 species (41.8%) constitute the freshwater Mastigophora, Rhizopoda and Ciliophora, respectively (Table 2.1). Considering 4,100 ciliate species, Finlay and Esteban (1998) grouped 2,049 (or 52%), 1,578 (or 38%) and 473 (or < 12%) species as marine, freshwater and terrestrial inhabitants, respectively. Incidentally, the estimated 12% for the terrestrial ciliates is comparable to that of Foissner (2014). *Of 26,450 species, 73.9, 19.9 and 6.2% of the three classes of protozoans inhabit the marine, freshwater and terrestrial habitat, respectively. The number and proportion of freshwater protozoans decrease in the following descending order: Ciliophora < Mastigophora < Rhizopoda.*

TABLE 2.2

Estimation on spatial distribution on the number of parasitic/symbiotic protozoan species on hosts from different habitats

Class and Group	Marine	Freshwater	Terrestrial	Total
		Mastigophora		
Symbionts (Table 4.3)	–	–	> 263	1900
Parasites (Table 1.2)	–	–	< 1637	
		Rhizopoda		
Entamoeba spp (Fig. 1.10B)	–	–	24	
Piroplasmea (Table 1.3)	–	–	75	250
Proteomyxidia† (Table 1.3)	–	62	62	
Sporozoa (Table 1.10)	3710	718	2071	6500
		Ciliophora		
Symbionts (Table 4.4)	–	–	> 500	2500
Parasites (Tables 2.1, 4.6)	1145*	1355	–	
Total	**4855, 41.7%**	**2135, 18.3%**	**4632, 39.9%**	**11150**

† considering the distribution on algae and crop plants, * based on 48.8% for marine (Table 2.1)

Except for cysts voided on terrestrial soil (e.g. *Eimeria*), almost all parasitic protozoans within their hosts are surrounded by an 'aquatic environ', irrespective of their habitats. However, a survey was made to relate the protozoan parasites/symbionts and their host's habitat (Table 2.2). *Of 11,150 protozoans, 42, 18 and 48% are parasitic/symbiotic species that are hosted by marine, freshwater and terrestrial metazoan hosts, respectively.* Comparative values between the spatial distribution of metazoans, plants and protozoans are listed below: Strikingly, only 6% free-living protozoans have gained access to terrestrial habitats, in comparison to 88% and 77% of plants and metazoans. Incidentally, Foissner (2014) also reported that of 2.7 g/m^2 biomass of soil inhabiting animals, 31% is contributed by protozoans; the testate rhizopods requires ~ 22 days to divide, in comparison to a few hours for the aquatic protozoans (see p 118). Soil porosity is a decisively important factor on incidence and abundance of free-living terrestrial protozoans. In pores smaller than 0.8 µm, soil/sediment-inhabiting bacteria do not exist. Consequently, bacteria-feeding protozoans are also not found in soil/sediment with porosity of < 1 µm (Wang et al., 2005).

	Marine	**Freshwater**	**Terrestrial**
Plants (%)	4.8	7.2	88.0
	12% aquatic + 88% terrestrial		
Metazoa (%)	15.1	7.8	77.1
	22.9% aquatic + 77.1 % terrestrial		
Plants + Metazoa (%)	17.5% aquatic + 82.5% terrestrial		
Free-living Protozoa (%)	74.0	20.0	06.0
	94.0% aquatic + 6.0% terrestrial		
For all protozoans (%)	71.0	18.0	11.0
	88.0% aquatic + 11.0% terrestrial		

2.2 Colonization of Land

Before explaining the said contrasting spatial distribution, a sidewalk is required. Life has existed in the sea longer than on land. Fossils reveal the existence of bacteria over 3.7 and 3.1 billion years ago (BYA) in the oceans and land, respectively. Interestingly, the greater variety of environmental niches in land provides a better scope for specialization, which has led to denser occupation of terrestrial space and in turn, to more biological niches. Species richness and diversity on land seems to have overtaken that in the oceans ~ 125 million years ago (MYA) (Costello and Chaudhary, 2017). Of 3,300 speciose cyanobionts, some 689 species or 21% inhabit the moist rocks and damp soils. For example, the filamentous heterocystous (nitrogen fixing cells) genus *Scytonema* alone is composed of 168 species. So are the 21,925 speciose bryophytes. Unlike vascular plants, these bryophytes lack a cuticular barrier and are characterized by large scale exchange of cations, i.e. they can absorb water and nutrients through the entire body surface, in addition to their photosynthetic capacity and engaging cyanobionts with nitrogen fixing capacity.

Irrespective of structural simplicity, autotrophism seems to have allowed the cyanobionts and bryophytes to occupy land, albeit relegated to moist zones alone. Protozoa, known to have originated some 1.5 BYA, are also as porous as bryophytes are. As saprozoics, they also acquire micronutrients through their body surface (see Chapter 3). However, the combination of the structural simplicity and heterotrophism in Protozoa seems to have hindered their colonization of land. In fact, the structurally simpler bryophytes, originated during the Silurian Era (425 MYA), had to wait until the arrival of the structurally well-organized angiosperms during the Jurassic Era (222 MYA) to effectively colonize the land and explosively diversify during the Cretaceous Era (see Pandian, 2022). So are the metazoans. Though abundant during the Cambrian waters (600 MYA), they also had to await the dawn of < 450 speciose earthworms (see Pandian, 2021b) during the Jurassic (209 MYA, *blogs.biomedcentral.com*) and the arrival of structurally well-organized arthropods to conquer and colonize the entire spectrum of land area during the Jurassic – Cretaceous Eras (225 MYA) and to undergo explosive diversification (Pandian, 2021b).

Of 12,012 digenean platyhelminthic species, ~ 5,500 species or 46% digeneans have 'migrated' into land by parasitizing some 33,000 speciose terrestrial vertebrates. So are the 2,821 speciose taenioid cestodes (see Pandian, 2020). This holds true also for 11,150 protozoan species or 33.6% of protozoan parasites and symbionts (Table 2.2, see also Chapter 4) in terrestrial invertebrates and vertebrates. However, only 1,630 free-living species or 6% protozoans could occupy moist and damp soils in land. Notably, it is not motility that has hindered the colonization of land, as terrestrial plants are also not motile. *Irrespective of motile or non-motile, autotrophism or heterotrophism, it is the complex structural organization of higher plants and metazoans that has contributed to the conquest and colonization of land; the combination of structural simplicity and heterotrophism of free-living protozoans have hindered the colonization of land*. Hence, migration to the terrestrial habitat through parasitization seems to be a common feature.

3

Acquisition of Nutrients and Food

Introduction

In trophic dynamics, the significant role played by protozoans has yet not been adequately recognized. The following may provide an idea on their key role in trophic dynamics. Firstly, the flagellates (15%) and ciliates (32%) constitute 20% of zooplankton biomass (Finlay and Esteban, 1998). In sediments too, their biomass can be high. In terms of volume, their density in sediment is almost equal to that in water column, i.e. 1 to 10,000 cell/cm^3 (see Finlay and Esteban, 1998). The ciliate biomass in the sediment can be 23 kg dry weight/ha (Finlay and Esteban, 1998). Secondly, their doubling time is also almost equal to a few bacterial species, on which they feed (Table 3.1). The inoculation of 10 bacterial cells may produce up to 1,000 million within 17–24 hours. The flagellates ingest a few to 100 bacterium/flagellate/h and some ciliates ingest > 1,000 bacterium/ciliate/h (Sanders, 2009). Consequently, bacterial growth is followed and contained by grazing, for example, by flagellates (Fig. 3.1A) and nanoflagellates (Fig. 3.1B). In the course of 200 hours in the former and 80 hours in the latter, the bacterial biomass is contained by the

TABLE 3.1

Doubling time of some bacteria (Gibson et al., 2018) and protozoans. *Limnohabitans*, Hyman (1940)[‡], Hawes (1963)[⁑], Wang et al. (2005)[†], Lekfeldt and Ronn (2008)[♣], Grujcic et al. (2015)[*]

Species	Doubling time (h)	Species	Doubling time (h)
Bacteria			
Vibrio cholerae	1.1	*Escherichia coli*	15.0
Staphylococcus aureus	1.9	*Salmonella enterica*	25.0
Pseudomonas aeruginosa	2.3		
Protozoa			
Ciliates		*Keronopsis*[†]	56
Colpodium[†]	6	*Paramecium aurelia*[‡]	24
Uronema[†]	7	Flagellates[*]	10
Euplotes plicatum[†]	22	*Cercomonas*[♣]	24
E. vannus[†]	26	*Amoeba proteus*[⁑]	22

heterotrophic flagellates. In fact, the protozoan feeding on bacteria may increase the turnover rate of essential nutrients that would otherwise remain 'locked up' in bacterial biomass (e.g. Fenchel and Harrison, 1976). From laboratory experiments, Laybourn and Stewart (1975) brought evidence for increasing bacteria feeding by protozoans, as the ratio of the bacteria to protozoa is increased (Fig. 3.1C). In fact, chemical cues appearing from *Listeria* induce the migration of *Euglena* toward the bacteria (Gaines et al., 2019). Hence, grazing by protozoans stimulates the decomposition rate of organic matter. Not surprisingly, the turnover rate of food biomass into consumer biomass is nearly twice as fast for protozoans than that for metazoans. For example, the productivity rate of ciliates is 564 kg/ha/y, in comparison to 290 kg/ha/y for metazoans (see Finlay and Esteban, 1998). Thirdly, protozoans account for ~ 66% bactérivory in plankton (Carrias et al., 1996). The ciliates can capture bacteria as small as 0.3–1.5 µm and transfer 94% bacteria (e.g. from the effluence of man-made wastewater treatment plants, Horan, 2003) to the next trophic level; otherwise, these bacteria may remain as uncapturable biomass. *In terms of biomass, short (doubling) generation time and ability to capture microscopic bacteria and make them available to macroscopic consumers, the protozoans play a key role in trophic dynamics.*

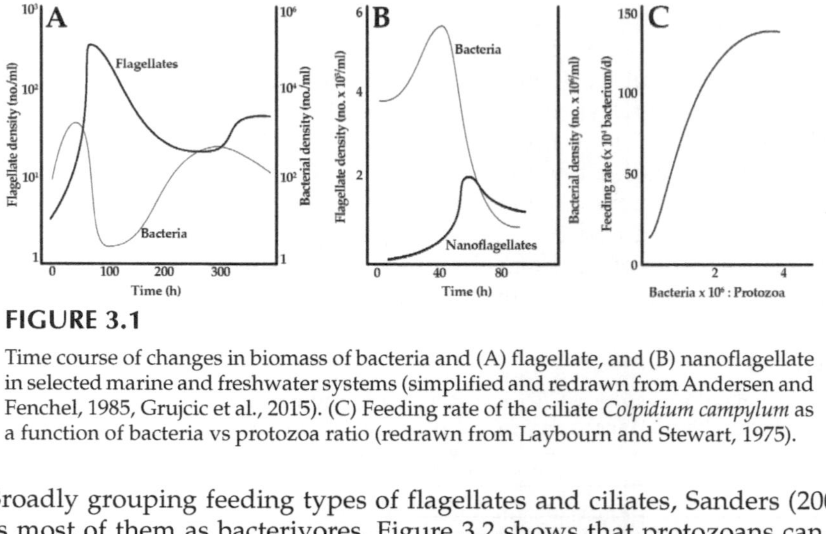

FIGURE 3.1

Time course of changes in biomass of bacteria and (A) flagellate, and (B) nanoflagellate in selected marine and freshwater systems (simplified and redrawn from Andersen and Fenchel, 1985, Grujcic et al., 2015). (C) Feeding rate of the ciliate *Colpidium campylum* as a function of bacteria vs protozoa ratio (redrawn from Laybourn and Stewart, 1975).

Broadly grouping feeding types of flagellates and ciliates, Sanders (2009) lists most of them as bacterivores. Figure 3.2 shows that protozoans can be bacterivores (e.g. *Rhynchomonas, Pteridomonas, Monosiga, Hexamita, Plagiopyla,* Fig. 3.2A–D, K) or algivores (e.g. *Protoperidinium, Tintinnopsis, Coleps,* Fig. 3.2E, N, O) or carnivores (e.g. *Amoeba, Actinophrys* [can predate rotifers also], *Heliosphaera,* Fig. 3.2H–J) or detrivores (e.g. flagellate: *Paramnema,* amoebae: *Mayorella,* ciliates: *Prorodon, Urotricha,* Sanders, 2009). It should, however, be noted that the flagellates may capture bacteria by their sticky flagella, which generate water currents in different directions (Fig. 3.2P–R), while the ciliates can either be true filter-feeders or predators. When the prey size (length) is less than tenth the size of predator, filter-feeding prevails. But when it exceeds a tenth of a predator size, the ciliates switch to predatory feeding (Fenchel, 1986). At the bacterial density of 10^5–10^6/ml, the ciliates, for example, are predatory but they are filter-feeders at higher (10^8/ml) bacterial densities (Fenchel, 1980a). Hence, *it is the combination of prey size and density that determines filter-feeding or predatory feeding in protozoans, especially ciliates.*

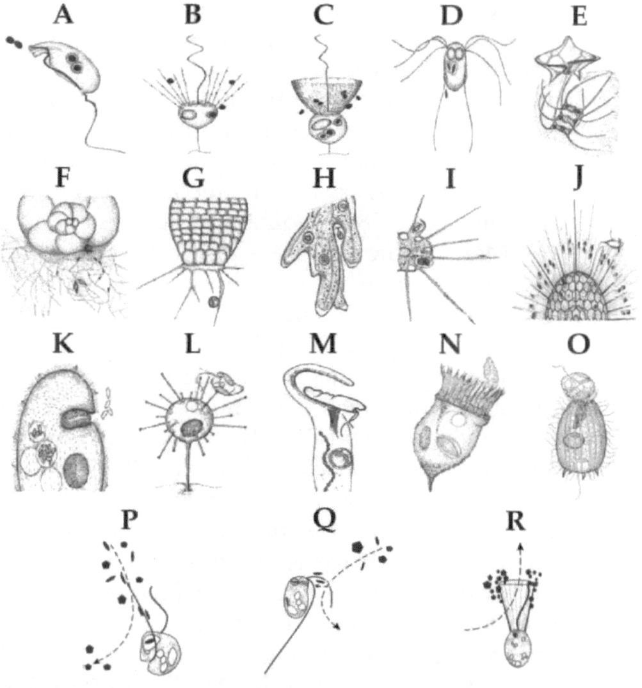

FIGURE 3.2

Upper row: Bacteria-feeding flagellates: (A) bodonid *Rhynchomonas*, (B) heliozoan *Pteridomonas*, (C) choanoflagellate *Monosiga*, (D) anaerobic diplomonad *Hexamita* and (E) heterotrophic dinoflagellate *Protoperidinium* trapping a diatom. Middle row: Rhizopods: (F) bacteria and a diatom being trapped by the foraminifer *Rotalia*, (G) a cell being stuck in the sticky tilopod of testate amoeba *Assulina*, (H) pseudopodial trapping of a flagellate in the naked *Amoeba*, (I) ingestion of a flagellate by the heliozoan *Actinophrys* and (J) radiolarian *Heliosphaera* with a large number of symbiotic dinoflagellates and a trapped tintinnid ciliate. Lower row: (K) bacteria being captured by anaerobic ciliate *Plagiopyla*, (L) diffusion-feeding suctorian *Podophyra* with a flagellate trapped on a feeding tentacle, (M) haptorid ciliate *Dileptus* with toxicysts used to kill the flagellate prey, (N) loricate tintinnid *Tintinnopsis* ingesting a diatom and (O) the gymnostome ciliate *Coleps* ingesting a dinoflagellate (modified and redrawn from Finlay and Esteban, 2017). Directions of flagellary current that enable their capturing selected bacteria and rejecting clay particles in the flagellate (P) *Spumella* sp, (Q) *Bodo* and (R) *Entosiphon* (free hand drawings from Boenigk and Novarino, 2004).

3.1 Feeding Types and Quantification

Based on the modes of nutrients and/or food acquisition, Protozoa may broadly be grouped into (i) autotrophs and (ii) heterotrophs. The latter comprise (a) fluid-feeders (e.g. osmophages), (b) microphages (suspension-feeders) and (c) macrophages. However, the boundaries between these groups are not distinctive or strict. For example, the volvocids and chloromonadids among the Phytomastigophora are completely autotrophs (Table 1.2). The cryptomonadids can be either autotrophs or heterotrophs. Many dinoflagellates can be heterotrophs and/or mixotrophs. At least, half of the 2,000 speciose dinoflagellates are exclusively heterotrophs and the other half combines heterotrophy with autotrophy (Stoecker, 1999). For example, the dinoflagellates *Procentrum minimum* and *Scrippsiella trochoidea* are mixotrophs and constitute 98% biomass in the Alexandria Harbor, Egypt (Ismael, 2003). Many protozoans are also designated as saprozoic, i.e. acquisition of micronutrients from the surrounding water through the body surface. There are many radiolarians colonial species that engage symbiotic green algae and are partially heterotrophs and autotrophs (see Decelle et al., 2013). Similarly, the symbiont-bearing foraminifers are also partially heterotrophs and autotrophs (see Hallock, 1999). There are other heterotrophs, who partially depend on Dissolved Organic Matter (DOM) (e.g. the foraminifer *Notodendrodes antarctikos*, DeLaca, 1982). These idiosyncratic combinations of autotrophy/mixotrophy, autotrophy/heterotrophy and heterotrophy/osmotrophy necessitate some approximations/compromises to enable the quantification of protozoans into one or other group.

Before recording the quantification aspects, a few sentences should be introduced regarding DOM. One liter of surface sea water contains 1 mg total DOM. Free amino acids, comprising 5% of DOM, are dissolved at 5×10^{-7} M/l concentration in free water and 1.1×10^{-4} M/l in the interstitial sediment water (for more details, see Pandian, 1975). Many protozoans occur and abound in the sediment waters. Not surprisingly, Putter (1909) rightly claimed that DOM may also be absorbed across the body surface and used as a nutrient source by animals including protozoans. An argument against Putter's claim is that if DOM can be absorbed per se, the DOM from body fluids can also be leaked through the body surface into ambient water. Ferguson (1971, 1972) measured both influx and efflux of free amino acids in many invertebrates and found that the net influx of amino acids is overwhelmingly inward. Thus, the amino acid uptake alone contributes up to 25% of the nutrient requirements in an enchytraeid worm and possibly up to 20% in another polychaete species (see Pandian, 2019). Hence, the possibility does exist for many protozoans to acquire DOM from the ambient water through their 'porous' body surface, and partially or wholly satisfy their nutritional requirement. While it may be costlier for free-living animals to acquire DOM against osmotic gradient from the ambient water (e.g. gutless oligochaetes), the passive acquisition of required micronutrients by parasitic protozoans from the surrounding fluid medium of the host can be rapid and less costly (see Pandian, 2020, 2021b). Among metazoans, there are leeches, bugs and mosquitoes that can actively suck the body fluids of their hosts but the protozoans are unable to do it, albeit diffusion feeding is reported for a few ciliates.

Within the microphagous suspension feeders, two groups are recognized: (i) filter-feeders turn over the relatively less dense water medium to filter and acquire food particles and (ii) sediment/deposit-feeders turn over the denser sediment substratum. The latter is costlier and requires relatively more investment than that in the former. Not surprisingly, the filter-feeders constitute 3.2% of all metazoans, in comparison to 1.5% for the sediment-feeders (see Pandian, 2021b). It may be very costly for protozoans to acquire nutrients as sediment feeders. In fact, many anaerobic protozoans actively graze over the sediment but not deep within the sediment. Being a rarity, only a three foraminifer species are 'wrongly' considered by Goldstein (1999) as deposit feeders; they are deep sea dwelling *Globobulimina pacifica*, *Uvigerina peregrina* and shallow water *Ammonia beccarii*. They feed more on the bacteria and detritus on sediment surface rather than the sediment itself. On the whole, protozoans are considered as incapable of sediment/deposit-feeding, as three species out of 32,950 or 0.01% alone are indicated to feed as "deposit-feeders" by Goldstein (1999). Secondly, within filter-feeders, there are two subtypes: (i) sessile filter-feeders and (ii) motile filter-feeders. Among metazoans, the former includes sessiles like the bivalves, bryozoans and others, and the latter the rotifers, pelagic urochordates (e.g. Larvaceae, Salpidae, Doliolidae, see Pandian, 2018). The sessile filter-feeders consist of as many as 41,403 species, whereas the motile filter-feeders 2,902 species alone. Pandian (2021b) noted that it is costlier to be motile filter-feeders, and evolution deters their species diversity.

Through the following procedure, the number of species adopting different modes of food acquisition was reached (Table 3.2). (i) From Table 1.2, the number for the autotrophic protozoan was collected as ~ 1,600 species. (ii) For osmotrophism, the value of 11,150 parasitic species was drawn from Table 4.6. (iii) Regarding microphages or suspension feeders, the following values were obtained: 125 species for choanoflagellates, and 2,293 species for sessile filter-feeding ciliates and 2,675 species for motile filter-feeding ciliates, as indicated below: Horan (2003) indicated that the Holotrichia and Spirotrichia are motile ciliates with well-developed structures for filter feeding. But the Peritrichia and Suctoria are sessile filter-feeders.

From computer search (*onezoom.org*), the following species numbers were arrived: the values reported in Table 1.12 may be underestimates.

Motile filter-feeders (species no.)		Sessile filter-feeders (species no.)	
Holotrichia	160	Peritrichia	1,837
Spirotrichia	2,515	Suctoria	456
Subtotal	2,675	Subtotal	2,293
Total number of filter-feeders: 4,960 species			

(iv) The value of 15,107 species or 46.2% of all protozoans was reached for macrophagy, after subtracting 17,743 species estimated for autotrophy, osmotrophy and microphagy (Table 3.2).

TABLE 3.2

Estimation on the number of species adopting different modes of food acquisition

Modes of food acquisition		Species	
		(no.)	(%)
1. **Autotrophism** (see Table 1.2) Volvocids, Chloromonadids, Coccolithophores, 50% of dinoflagellates* (Stoecker, 1999)*		**1,600**	**4.9**
2. **Osmotrophism** (see Table 4.6, all parasites)			
a. Mastigophora	1,900		
b. Rhizopoda	250		
c. Sporozoa	6,500		
d. Ciliophora	2,500		
	Subtotal	**11,150**	**33.8**
3. **Microphagy** (suspension feeders)			
a. Sessiles			
(i) Choanoflagellates	125		
(ii) Ciliates	2,293		
Subtotal	2,418 or 7.3%		
b. Motiles	2,675 or 8.1%		
	Subtotal	**5093**	**15.4**
4. **Macrophagy*** 1,600 + 11,150 + 5,093 = 17,843. 32,950 – 17,843 = 15,107			**45.8**

A comparison of the feeding modes between protozoans and metazoans suggests that (i) *the structural simplicity of protozoans has not facilitated the development of (a) feeding by fluid sucking, (b) osmotrophism in free-living species and (c) deposit feeding* (Table 3.3). *But it has let the manifestation of an unusually high proportion (33.8%) parasitism and filter-feeding (15.4%), especially in ciliates.*

Understandably, the simplicity has led to low proportion (46%) predatory feeding, in comparison to 77% in the structurally well-developed metazoans. Incidentally, of ~ 1,500 ciliate filter-feeders from the sediments of lakes and rivers, there are only 25.32 ciliate species, although the number of ciliate species is increased almost linearly with increasing number of ciliates (Fig. 3.3). Clearly, *the ciliates have gone more for numerical diversity rather than species diversity.*

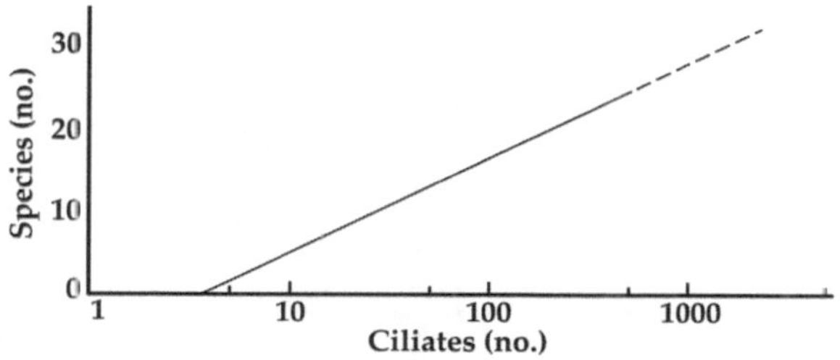

FIGURE 3.3

A representative example for species diversity with increasing number of free-living ciliates in sediments from lakes and rivers (modified and redrawn from Finlay and Esteban, 1998).

TABLE 3.3

A comparative account on the presence and absence of different modes of food acquisition in protozoans and metazoans

		Metazoans		Protozoans
I. **Autotrophy**			0%	4.9%
IIa. Fluid-feeders				
Fluid-suckers				
	i. Free-living	0.72%		0%
	ii. Pests	10.65%		0%
	iii. Parasites	2.8%		0%
		Subtotal: 14.2%		0%
IIb. Osmotrophy				
i.	Free-living	0.006%		
ii.	Parasites	0.6%		11,150
		Subtotal : 0.61%		33.8%
III. **Microphagy**: Suspension-feeders				
i.	Filter-feeders	3.2%	6.6%	15.4%
ii.	Sediment-feeders	1.7%		0%
	Aquatic	1.5%		0%
	Terrestrial	0.2%		0%
IV. **Macrophagy**				
			76.6%	45.8%

3.2 The Filter-Feeding Ciliates

Thanks to Dr. Tom Fenchel, fairly adequate information is available on filter-feeding ciliates. For his contributions, Fenchel (1980a, b) selected ~ 20 ciliate species. Surprisingly, *all 20 are motile filter-feeders*. For metazoans, available information is mostly on sessile filter-feeding (see Pandian, 1975). In fact, Fenchel's contributions on the motile filter-feeding represent a complement and another dimension to have a holistic view on filter-feeding in animals. The following briefly summarizes Fenchel's contributions: 1. With increasing body length, the body size, as measured in volume, increases linearly in 14 ciliate species (Fig. 3.4A). 2. Expectedly, the ingestion rate decreases with

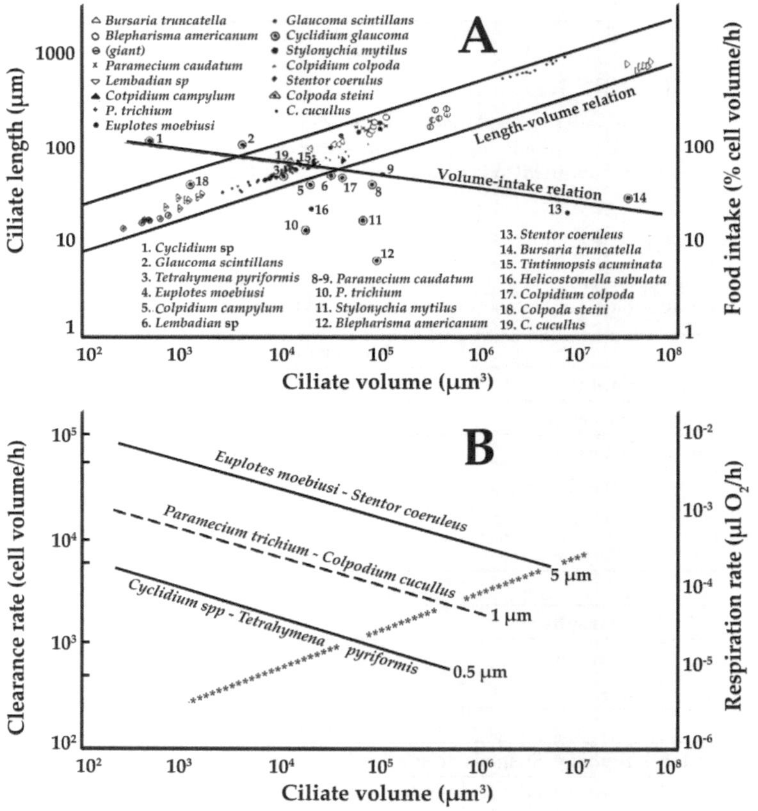

FIGURE 3.4

Ciliates: (A) Body length as a function of volume in 14 species. Food intake rate as function of volume in 19 species. (B) Clearance rate as function of ciliate size. Note the parallel trend levels with decreasing size in three groups (compiled and redrawn from Fenchel, 1980a). Respiration rate (*****) as a function of ciliate volume at 20°C (redrawn from Laybourn and Finlay, 1976).

increasing body size in 19 ciliate species (Fig. 3.4A). 3. With increasing ciliate's size, the volume of water filtered, i.e. clearance rate decreases. The trends are size specific; the one for the ciliate size – clearance rate is at a higher level for the largest ciliates of 5 µm group than that for the smallest size group of 0.5 µm (Fig. 3.4B). 4. With increasing clearance rate, energy required to filter increasing volume of water, as indicated by respiration rate linearly increases in the ciliates.

5. With increasing (particle) food size, food intake by filter-feeding decreases, especially beyond 10 µm size, where the ciliates tend to either retain filter-feeding or switch over to predatory feeding (as shown by dotted lines in Fig. 3.5A). 6. With increasing (particle) food size, the amount of food ingested increases, as represented by *Colpidium campylum* (Fig. 3.5B). 7. Some of these relationships could be a little modified with the selection of prey size (a) with increasing prey size from 0.1 µm to 1.0 µm, the small sized holotrichid ciliates increases the clearance rate up to a maximum of 3 µl/h but up to 7.5 µl/h in the larger sized spirotrichid ciliates (Fig. 3.6). Extending his findings to the open ocean and sediments, as well as littoral zone and eutrophic lakes, Fenchel (1980a) generalized the following: 1. The ability of protozoans, especially ciliates to gather and concentrate the microscopic bacterial and other prey is equal or even excels that of metazoans. 2. With number < 5–7/10^5 bacterium/ml or biomass of < 6×10^4 µm^3/ml, the open oceans may not support filter-feeding ciliates. Conversely, interstitial waters in sediments, and waters in eutrophicated lakes with > 10^8 prey/ml can sustain them.

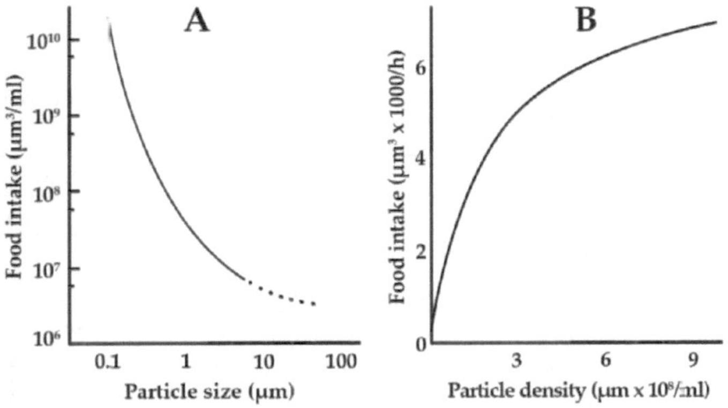

FIGURE 3.5

(A) Food intake as a function of particle size in 13 ciliate species. (B) Food intake as a function of particle density in *Colpidium campylum* (modified and redrawn from Fenchel, 1980a, b).

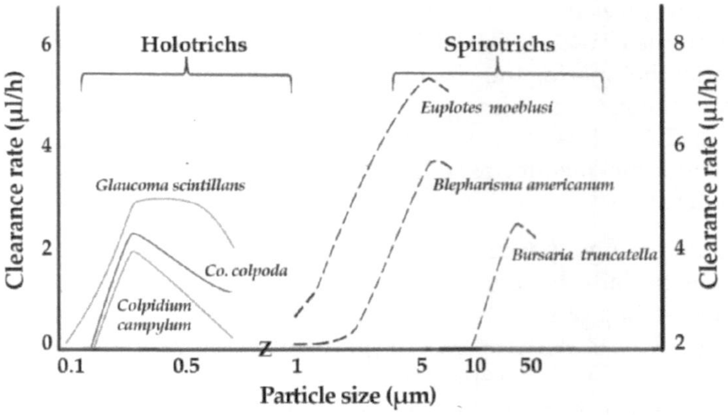

FIGURE 3.6

Particle selection as measured by clearance rate in selected holotrichs (———) and spirotrichs (- - -) (compiled from Fenchel, 1980b).

Another aspect that merits consideration is (i) the minimum range of (particle) food sizes that are retained by the filter-feeding systems in protozoans and metazoans and (ii) the respiratory or acquisition cost to gather and ingest food particles, as measured by a clearance rate (Table 3.4). Protozoans, especially ciliates can retain food particles of the size ranging from 0.3–1.5 µm, as in *Colpoda* to 20–60 µm, as in *Bursaria* among the motile filter-feeders. The sessile ciliates may also retain food particles of a similar size range, albeit no information is yet available for them. The corresponding

TABLE 3.4

Comparative account on the particle size filtered and O_2 consumption in sessile and motile filter-feeders (compiled from Jorgensen, 1966, Pandian, 1975, 2021b, Fenchel, 1980a)

Species	Food size (µm)	Mean (µm)	Species	Food size (µm)	Mean (µm)
Protozoa			**Metazoa**		
Mobile filter-feeders					
13 ciliate spp	0.2–10.0	5.10	Cladocera	1–17	9.0
Colpoda	0.3–1.5	0.90	Rotifera	4–17	10.5
Holotrichs	0.5–1.0	0.75	Larvae of		
Paramecium	2.0–5.7	3.85	Nudibranchs	10–15	12.5
Spirotrichs	4.0–50.0	27.00	Echinoderms	65–200	132.5
Bursaria	20.0–60.0	40.00			
	X̄ of means	**12.9**		**X̄ of means**	**41.3**

Table 3.4 contd. ...

...Table 3.4 contd.

Species	**Food size**	**Mean**	**Species**	**Food size**	**Mean**
	(µm)			**(µm)**	
Protozoa			**Metazoa**		
Sessile filter-feeders					
No value is available			Sponges	1– 5	2.5
			Bivalves	1–5	2.5
$\bar{X}$ of means				**$\bar{X}$ of means**	**2.5**
Clearance rate (l/ml oxygen uptake)					
Motile filter-feeders					
Lambadion		0.15	Cladocera	1.3–2.6	3.9
Stentor		0.2	Bivalve larvae	5–15	10.0
Bursaria		0.6	Copepods	5–60	32.5
Paramecium		1	Echinoderm larvae	12–170	91.0
Colpidium		2	Tunicates	13–300	156.0
Cyclidium		3			
Glaucoma		8			
$\bar{X}$ of means		**2.14**		**$\bar{X}$ of means**	**58.7**
Sessile filter-feeders					
No value is available on computer search			Sponges	4–23	13.5
			Bivalves	4–79	41.5
			Bryozoa	12–60	36.0
				$\bar{X}$ of means	**30.3**

values for metazoans range from 1–17 µm for the Cladocera to 62–200 µm for the echinoderm larvae among motile filter-feeders. For sessile metazoans, these values range from 1 µm for sponges to 2.5 µm in bivalves. *The motile protozoans are able to filter food particles as small as 13 µm, in comparison to 41 µm by motile filter-feeding metazoans, i.e. the protozoans are able to retain and feed on 3.2 times smaller microbes and other preys, in comparison to those of metazoans. Clearly, protozoans can gather and ingest much smaller bacteria and others, and reduce their populations more effectively* than their metazoan counterparts.

The clerance rate ranges from 0.15 l/ml oxygen (O_2) uptake for motile protozoan like *Lambadion* to 8 l/ml O_2 uptake in *Glaucoma* (Table 3.4). The rate of 2.14 l/ml O_2 uptake represents the mean of means. The corresponding values for motile metazoans range from 3.9 l/ml O_2 uptake in cladocera to

156 l/ml O_2 uptake in tunicates and averages 58.7 l/ml O_2 uptake. Contrastingly, the mean value for sessile filter-feeding metazoans is 30.3 l/ml O_2 uptake, i.e. *among the metazoans, the cost of acquiring food through clearance rate is twice costlier for motile filter-feeders than that of sessile filter-feeders.* Incidentally, these values may considerably change, when clearance rates are considered on the basis of O_2 uptake in l/g body size/h. *Despite their structural simplicity, motile filter-feeding ciliates can not only filter to gather and feed on much smaller microscopic bacteria and other preys but also do it at the lowest respiratory/clearance cost. This may be a reason for the existence of 15.5% filter-feeding protozoans, in comparison to 3.2% filter-feeding metazoans* (Table 3.4). Perhaps, the 'handling cost' of preying bacteria is far less for ciliates than that of larger prey by metazoans.

4

Commensals—Symbionts–Parasites

Introduction

Among Protozoa, modes of life ranges (i) from autotrophism to mixotrophism and to heterotrophism, (ii) from motiles to sessiles, (iii) from solitaries to colonials and (iv) from free-living to commensalism, symbiosis and to parasitism. Various aspects on the first three have already been described. Hence, the present chapter is devoted to commensalism, symbiosis and parasitism in the context of species diversity. Commensals may survive without the host but the parasites cannot (see Jennings, 1977); their association can be beneficial only to parasites but not to the hosts, which suffer a competitive ability to survive, grow and reproduce, and in some cases cause death. On the other hand, symbionts cannot also survive in the absence of their respective partners but the association is mutually beneficial.

4.1 Commensalism

Commensals can be divided into two groups: (i) Ectocommensals can survive without the hosts and (ii) endocommensals can also survive without the hosts for a few days but not permanently (Jennings, 1977). The 54 speciose flagellated *Trichomonas* (*onezoom.org*) are commensals found in the gut of vertebrates, leeches and termites. They feed on bacteria and yeast available in the gut (see Hyman, 1940). The protozoan commensals are common mostly among ciliates. They can be considered under the following two groups: (A) On motile hosts, the stalked sessile commensals 'enjoy' free transport but have to acquire food on their own (e.g. ~ 500 species). (B) Utilizing the 'free of cost', ciliary current brought by the sessile bivalve and branchial chamber of fish hosts, commensal ciliates filter their own food. In other words, *sessile commensals enjoy free transport but acquire food at their cost, whereas the free-living commensals gain from the ciliary current of the host to filter their own food.* Table 4.1 lists 141 free-living endocommensals/facultative parasitic ciliate species. However, the boundaries remain open between the endocommensals and facultative parasitism.

TABLE 4.1

Estimation of number of commensal ciliates species on bivalves. Names of genera are identified by Lauckner (1983) and their species number are arrived from *onezoom.org, algaebase.org**

Genus	Species (no.)	Genus	Species (no.)
Endosphaera	3*	*Boveria*	9
Insignicoma	1	*Proboveria*	3
Gargarius	1	*Anoplophrya*	57
Uronema	31	*Ellobiophrya*	5
Conchophyllum	1	*Licnophora*	11
Thigmophrya	5	*Uronychia*	13
Peniculistoma	1	**Total**	**141**

4.2 Symbiosis

Algae – Protozoans: Available information on protozoan symbiosis is very widely scattered. Tedious computer searches have collected and collated relevant information. Perhaps, for the first time, the protozoan symbiosis is elaborated in this account. It is considered under two groups: Group 1 consists of symbiosis between algae and protozoans and Group 2 includes cellulase digesting subgroup 2a the flagellates in herbivorous insects and 2b ciliates in combination with bacteria and/or fungi in herbivorous mammals. Group 1 comprises ~ 344 host protozoan species and 14 algal symbiont species, as represented by cyanophytes, chlorophytes and rhodophytes and an unknown number of dinoflagellate species (Table 4.2). Maltose is perhaps the only photosynthate released by these intracellular symbionts, but it is maltose and glucose by *Chlorohydra viridissima*, galactose and glucose by the symbiont to the molluscan host and glycerol by *Symbiodinium* to the cnidarian host (see Pandian, 1975). The 344 host species consist of rhizopodan radiolarians and foraminifers and one ciliate species. When kept in the dark, *Paramecium bursaria* dies. But in the presence of light, the symbiotic algae supply all the required nutrients (except for some salts) to *P. bursaria* to survive, grow and reproduce (see Hyman, 1940).

Flagellates and herbivorous insects: The diversification and evolutionary success of herbivorous insects in a way depend on their myriad relationships of symbiotic protozoans and prokaryotes to utilize nutrient-poor feed and

TABLE 4.2

Estimation of species numbers in symbiotic algae and protozoan hosts

Symbiont	Host	Reported observations	Reference
Dinoflagellates	245 radiolarian species	CO_2 and nitrogen as ammonia released to symbionts, which provide a jelly-like layer to protect and food capturing by radiolarians	*ucmp.berkeley.-edu, eol.org*, Probert et al. (2014)
13 Photo-symbionts: diatom species, dinoflagellates, chlorophytes, rhodophytes and/ or cyanobacteria[†]	98 foraminiferan (56 globigerinid species + 40 loborotalid species + 2 candeinid species*	CO_2 and nitrogen released by host. Maltose is supplied by symbionts	Takagi et al. (2019)[‡], *onezoom. org*[*],Lee (2006)[†]
1 *Chlorella*	Ciliata: 1 *Paramecium bursaria*	*Paramecium* supplies CO_2, nitrogen and protection to hundreds of endosymbiotic algal cells. In return, *Chlorella* supplies maltose to its host	Bostick, M (*carolina.com*)
14+	**344**		

digest their recalcitrant components (Engel and Moran, 2013). The 350,000 speciose coleopterans and 160,000 speciose lepidopterans (see Pandian, 2021b) seem to depend entirely on microbes to digest the cellulosic soft leaves and other components of plants (see Engel and Moran, 2013). However, lignocellulose is the most prominent component of woody plants and dead plant materials. It is the most abundant biomass in terrestrial ecosystem and is degraded by the 2,600 speciose isopterans, many roaches within the 4,500 speciose Blattodea and a few crickets within the 20,000 speciose Orthoptera (see Pandian, 2021b for species number). The isopteran termites are important soil insects that efficiently decompose lignocellulose aided by protozoans and their endosymbiotic microbes. They can dissimilate a significant fraction (74 to 99%) of cellulose and (65 to 87%) hemicellulose (Ohkuma, 2003). Incidentally, cellulose is composed of glucose units polimerized together by β,1-4-glycosidic linkages, whereas hemicellulose consists of primarily xyloses linked together by β,1-4-xylosidic linkages. Pectin is a polymer of α,1-4 linked galacturonic acid units (see Dehority, 2002). Among Blattodea, the cockroaches are versatile omnivores and feed virtually any organic material including relatively refractive nitrogen-poor feeds such as plant cellulose residues (Gijzen and Barugahare, 1992). Both the termites (e.g. *Reticulitermes flavipes*) and roaches belong to the genus *Cryptocercus* feed strictly on wood and harbor flagellate protozoans in their hindgut, where they are hydrolyzed to wood polysaccharides (Dehority, 2002). On deprivation of the symbiotic flagellates, they starve and die (see

Hyman, 1940). Hence, these symbiotic associations are obligatory and species specific (Noda et al., 2009). The gregariousness of solitary roaches and crickets, and coloniality of termites ensure the reacquisition of the symbionts mostly through feces after every molt as well as by the benign offspring.

With regard to the engagement of symbiotic microbes, the herbivorous insects may be considered under three groups. Group 1: The roaches may engage flagellates belonging to 17 genera in 10 families, as indicated by the sample of 31 taxa of bacteroid microbes (Ohkuma et al., 2009). The 2,600 speciose (see Pandian, 2021b) Isoptera comprise a complex assemblage of diverse species, roughly divided into 1,950 speciose higher termites belonging to the family Termitidae and 650 speciose lower termites (Abe et al., 2000). Group 2: The lower termites engage flagellates as well as microbes. Group 3: The higher termites harbor a dense and diverse array of prokaryotes, basidiomycete fungi cultivated in their nest but typically do not harbor flagellated protozoans. Besides, they may have their own cellulase (Ohkuma, 2003). *On the whole, ~ 700 herbivorous insects may symbiotically engage ~ 270 flagellate species* (Table 4.3).

TABLE 4.3

Estimation of symbiotic flagellate species in herbivorous invertebrates. The number of species is arrived from *onezoom.org**, *algaebase.org*†, *Wikipedia***, *gni.globalnews.org*‡

Genus	Species (no.)	Genus	Species (no.)
Barbulonympha		*Protrichonympha*	
Caduceisa		*Pseudotrichonympha*	8*
Cryptocercus	12**	*Pyrsonympha*	2*
Devescovina	28†	*Reticulitermes*	138*
Dinenympha	6*	*Rhynchonympha*	
Holomastigotoides	3*	*Snyderella*	4*
Hoplonympha		*Staurojoenina*	2*
Janickiella	2‡	*Stephanonympha*	4*
Joenia	3*	*Streblomastix*	1*
Lophomonas	2*	*Trichomitus*	4*
Metadevescovina	17*	*Trichonympha*	24†
Mixotricha	1*	*Urinympha*	1*
Oxymonas	1*	**Total**	**263+**

Gijzen and Barugahare (1992) reported an exceptional symbiotic digestion by the ciliate *Nyctotherus ovalis* in the roach *Periplaneta americana* (however, see also Vd'acny et al., 2018). The high density of ciliates in such large numbers (5 to 6 × 10^4/ml) in the hindgut suggests the key role played by ciliates. Reared on the cilia-free diet, the roach extends generation time, loses body weight and produces no methane. Hence, the cellulolytic ciliates and their endosymbiotic methanogenic bacteria are obligately required by the roach to maintain their normal generation time, body weight and methane production. This leads to the methane generation and ciliate-aided cellulose digestion in vertebrates.

Ciliates and herbivorous vertebrates: A preamble is necessary to describe the hereunder listed structural and other features of herbivorous vertebrates:

Fishes and reptiles
Anaerobic microorganisms present in chambers of the hindgut
Birds
Foregut, i.e. the gizzard, especially in the leaf eating *Opisthocomus hoazin*
Mammals: Hindgut
(a) Caecum fermenters, e.g. Rabbit, Guinea pig, Tapir (b) Colon fermenters, e.g. Horse, Ass, Zebra, Rhinoceros, Elephant. In horses and ponies, the proportion of cellulolytic bacteria ranges from 0.16 to 1.4% of total bacterial biomass
Mammals: Foregut
Orderly and synchronized contraction of the rumen and reticulum aids the regurgitation and chewing. Consequently, the particle size of the semi-digested feed that leaves the rumen to abomasum is 2 mm in sheep and 2 to 5 mm in cattle. Depending on the feed (dry or wet) type, saliva, rich in recycled urea and others, is secreted at the rate of 5 to 15 l/d in sheep and 75 to 190 l/d in cattle. Being the most important digestive organ, the rumen occupies three quarters of the abdominal space. It is indeed a large fermentation vat with the capacity of 3 to 15 l in sheep and 35 to 100 l in cattle. For more details, Dehority (2002) may be consulted.

The herbivorous vertebrates use symbiotic ciliates, bacteria and fungi. These symbionts are responsible for digestion of fibrous food to soluble sugars, which are subsequently fermented to yield Short Chain Fatty Acids (SCFAs), mostly in the form of acetate, propionate and butyrate, which provide 60–70% of the daily energy requirements, as in horses (Belzecki et al., 2015). However, complicated by recycling of urea through saliva, the interactions between the symbionts are far more complicated in ruminants (Newbold et al., 2015). In its hindgut, a horse can harbor 10^2 fungal cell/ml digestive fluid, 10^4 protozoan cell/ml and 10^5 archaean cell/ml and 10^{10} bacterial cell/ml. Being large in size, Protozoa constitute 50% biomass of the symbiotic biota (Belzecki et al., 2015). On the other hand, the ruminants hold far greater number of symbiotic biota. For example, the ciliate density ranges from 5.2×10^4 cell/ml to 44.5×10^4 cell/ml in domesticated cattle of Iceland, in comparison to the range from 130×10^4 cell/ml to 250×10^4 cell/ml in the wild reindeer. Remarkably, the number of symbiotic ciliates also ranges from 13 species + 2 subspecies in goat, to 19 species in wild reindeer, 27 species + 4 subspecies in sheep, and to 34 species + 10 subspecies in cattle of Iceland (Fuente et al., 2006). A large-scale survey aided by computer searches indicates that ~ 180 herbivorous mammals engage ~ 500 symbiotic ciliate species (Table 4.4).

TABLE 4.4

Estimation of symbiotic ciliates in vertebrate hosts

Family	Species (no.)	Reference
Ruminants: 164 species (*nhc.ed.ac.uk*)		
Trichostomatia	318	*onezoom.org*
Isotrichidae	2	*onezoom.org*
Ophryoscoletidae	< 48	*onezoom.org*
Cycloposthiidae	11	*onezoom.org*
Buctschliidae	9	*ncbi.nlm.nih.gov*
Balantiididae	> 77*	*onezoom.org*
Subtotal	465	
Perissodactyla: 16 species (*onezoom.org*)		
Blespharocorys	6	Belzecki et al. (2015)
Proboscidea: 7 species (*onezoom.org*)		
> 6 genera in 3 families	17	*onezoom.org*, Eloff and Hoven (1980)
Symbiont species number: 488, Host species number: 187		

77 species in the genus *Balantidium**

The interaction between rumen ciliates other members of microbiota is intense but greatly varied. Firstly, not all ruminant ciliates have hydrogenosomes to be only anaerobic (e.g. *Entodinium caudatum*, Firkins et al., 2020). Secondly, not all the ciliates carry the endosymbiotic methanogenic bacteria. The interaction between the rumen ciliates and members of the microbiota, though intense, is so varied that it is difficult to comprehend the specific role played by ciliates, bacteria and fungi. Rumen ciliates are represented by two morphologically and physiologically different groups: (a) the entodiniomorphids and (b) holotrichs (Michaiowski, 2005). 1. The ciliates such as *Polyplastron*, *Epidinium* and a few others belonging to the family Entodiniomorphida are cellulolytic and can be beneficial to ruminants feeding on fibrous foods/feeds. 2. The linear relationship between ciliate density and methane emission reveals that ~ 31% ruminant methane production may appear from the methanogenic endosymbionts carried by the ciliates. However, the sources are not yet known for the remaining 69% methane production by ruminants. This is an important area for research, as methane emission from the ruminants have to be reduced from polluting the atmosphere. 3. The rumen ciliates actively predate the bacteria, which serve as the main source of protein for them. Defaunation or deciliation renders the rumen to function more efficiently in terms of 30% more protein synthesis from the recycled urea by microbiota. Not surprisingly, defaunation improves the feed conversion ratio.

4.3 Parasitism

Australia has contributed 860 publications on Protozoa. Of them, 630 or 73% deal with parasitic species, whereas the remaining 230 or 27% are concerned with free-living species (Adlard and O'Donoghue, 1998). Among protozoans, an unusual proportion of 34% species are parasites and the remaining 66% are free-living species (see Table 4.6). Considering the Australian publications as an index, it is paradoxical to note that more information is available for the smaller number of parasitic protozoans than for the free-living species. In a way, it seems to be justified; for, the number of people and livestock risking infection, death due to infection and the overall loss or funds expended on their control are far higher (Table 4.5) than to those due helminths (see Pandian, 2019, 2020). Incidentally, values reported for fish are not specific, as those for man and livestock.

TABLE 4.5

Incidence and estimated cost or loss caused by protozoan parasites

Disease	Reported observations
Man	
Malaria	3.3 billion from 97 countries are at risk – 209 million are affected – 600,000 death – 3 billion US$ funded for malarial control – (WHO, 2015)
Trypanosomiasis	70 million from 36 sub-Saharan African countries are at risk (Simarro et al., 2012) – 6 to 7 million infection - up to 5 million deaths (*who.int*) – 624 million US$ loss (Lee et al., 2013)
Leishmaniasis	350 million are at risk – 12 million infected from 88 countries (Dr. Jorge Alvar, WHO Scientist)
Amoebiasis	50 million infection – 1 million deaths
Giardiasis	Loss of € 1.92 million
Trichomoniasis	3.7 million infections (*clevelandclinic.org*) – *Trichomonas fetus* – 26 million infection US alone (*businesswire.com*)
Livestock	
Trypanosomiasis	Cattle infection ranges from 4.3% in Burkina Faso to 35.7% in Uganda – 20% loss on productivity (Holt et al., 2016)
Trichomoniasis	~ 31% abortion – 192 US$ loss/cow in New Mexico alone (Wenzel et al., 2020)
Coccidiosis	Globally, 3 billion US$ poultry loss – net loss Rs. 3.4 billion in Central Java, Indonesia alone (Pawestri et al., 2019)
Fish	
	45% loss of flounder due to microsporidean *Glugea stephani* (Lom, 1984)
Dermosis	The disease caused by the coccid *Perkinnes marinis* kills ~ 30 million or 75% cultured oysters (Lauckner, 1983)

Many surveys have been made, regarding the number of parasite and host species and their incidence and intensity. A large-scale survey has revealed the distribution of free-living and parasite species among the four Superclasses of Protozoa (Table 4.6). The following generalizations can be noted: 1. Parasitism is limited to 1% in plants (see Pandian, 2022), and 7% in metazoans (see Pandian, 2021b), in comparison to 33.8% in protozoans. *The structural simplicity seems to have readily let the manifestation of parasitism in more numbers of Protozoa*; however, the structurally complex higher plants and animals do not let it. Interestingly, parasitism in crustaceans has secondarily simplified the hosts like *Sacculina carcini* (see Pandian, 1994). 2. Despite the lower efficiency of the food capturing by pseudopodia, only 2.2% rhizopods are parasites, as they seem to prefer symbiosis (see Table 4.2) rather than parasitism. With the lowest food capturing efficiency, a larger number of zoomastigophoran flagellates are driven toward parasitism (27.5%) rather than symbiosis (see Table 4.6). Clearly, it is *the structural simplicity of protozoans that has let the manifestation of parasitism rather than food capturing efficiency*.

TABLE 4.6

Estimated number of free-living and parasitic (including symbiotic) protozoan species (compiled from Levine et al., 1980, Adlard and O'Donoghue, 1998, Lynn, 2010a*, *parasite.org.au*†)

Superclass	Species (no.)	Free-living		Parasite	
		(no.)	(%)	(no.)	(%)
Mastigophora	6,900	5,100	73.9	1,900	27.5
Rhizopoda	11,550	11,300	97.8	250	2.2
Sporozoa	6,500†	0	0	6,500	100.0
Ciliophora	8000*	5,500	68.8	2,500	31.3
Total	32,950	21,900	66.5	11,150	33.8

Entry and exit: Ecto- and gut-parasites utilize the readily available easier entry and exit from the host, whereas the vascular parasites obligately require injective sanguinivorous vectors for transmission. The entry of the gut parasites may involve passive (e.g. *Nematopsis* into oyster) or active (e.g. ingestion of *Aggregata* from an infected crab by a benign *Sepia*) trophic mode of transmission. With regard to muscular parasites, the transmission also involves the trophic mode but the infected hosts have to be actively ingested by a benign host. Hence, the scope for entry and exit is of paramount importance from the point of transmission. In this context, the Protozoa may be considered under three categories: 1. Gut, 2. Muscular and 3. Vascular parasites.

In the absence of adhesive and/or clinging organelles, protozoan parasites can ill-afford ectoparasitism, albeit rare incidence like the flagellate *Cryptobia branchialis* on the flounder *Paralichthys dentatus* gills. They primarily inhabit the digestive tract and its associated organs. Considering fish as examples (Lom, 1984), their existence is known from the organs like the (i) liver (e.g. flagellate *Goussia clupearum* on *Clupea harengus, C. sparattus, Sardina pilchardus, Alosa sardina*, (ii) pancreas (e.g. coccidian *Calyptospora funduli* on *Fundulus grandi*, (iii) gall bladder (e.g. Myxosporidea *Ceratomyxa drepanosettae* on *Rheinhardtius hippoglossoides*), (iv) swim bladder (e.g. *Goussia caseosa* on *Macrourus berglax*) and (v) branchial chamber (e.g. *Haplosporidium* on oyster), as well as a little distant organs like the (vi) urinary bladder (e.g. myxosporid *Parvicapsula unicornis* on *Callionymus lyra*), (vii) testis (e.g. *Eimeria brevoortiana* on *Brevoortia tyrannus*) or (viii) ovary (e.g. microsporidid *Thelohania herediteria* on *Gammarus duebeni*). One way or the other, most of these associated organs are connected to the alimentary tract. Hence, they are grouped into the category, as the gut parasites.

The second category namely the vascular parasites consist of 156 speciose haemosporinans (e.g. *Plasmodium* spp), < 100 speciose trypanosomatines (e.g. *Trypanosoma, Leishmania*) and ~ 79 speciose rhizopodan piroplasmeans (*wikipedia, onezoom.org*, e.g. *Theileria equi*). In them, the injective mode of transmission by sanguinivorous vectors like the mosquitoes, tsetse fly, sandfly or ticks is obligately required. The 85 speciose haemogregarines are also corpuscular parasites. Their transmission involves ingestive but not injective mode; they are not included in the vascular category. Among the 1,100 speciose myxosporideans, ~ 633 species (*Hexacapsula* – 1, *Kudoa* – 117, *Myxidium* – 52, *Myxobolus* – 446, *Myxosoma* – 5, *Pentacapsula* – 4, *Unicapsula* – 8, *onezoom.org*) may be located within the host's muscles. *Unicapsula muscularis* in *Hippoglossus stenolepis*, *Kudoa histolytica* in *Scomber scombrus* and *K. musculoliquefaciens* in *Xiphias gladius* are some examples for this category.

Parasitic location: On the whole, the protozoan parasites consist of 11,150 species (Table 4.6). However, this survey recognizes that the life cycle of not more than 1,800 species, or < 20% parasites involve two hosts (see Table 4.9). In them, their parasitic sites differ; for example, *Trypanosoma* spp, *Theileria* spp and *Plasmodium* spp inhabit the gut tract of insects but on the blood corpuscles of vertebrates. These two hosted protozoan parasites complicate their categorization into the gut or vascular. Nevertheless, they are grouped into the vascular category from the point of their injective mode of transmission, while the others like the gregarines (e.g. *Nematopsis*) and eucoccids (e.g. *Aggregata*) are grouped into the gut category, as their transmission involves active and passive trophic modes alone. *Strikingly, almost all the two hosted protozoan parasites obligately involve one haemocoelomic hosts like the crabs, insects or leeches*. Firstly, it is easier to identify and quantify the number of species that can readily be brought under the vascular or muscular category. Secondly, the cumulative number of species belonging to these two categories (Table 4.6) was subtracted from the total 11,150 parasitic species number to reach at the number of species belonging to the gut parasitic category. Accordingly, *some 330 or 3.0% of parasitic species belong to the vascular category; ~ 633 or 5.6% of them belong to the muscular category. The remaining 10,187 or 91.4% protozoan parasites are located in the gut tract or its associated organs. With structural simplicity, a vast majority (91%) protozoan parasites have opted for their location on the gut tract and its associated organs.* As indicated earlier, there are no ecto-parasites among parasitic protozoans for want of adhesive and/or hanging organelles. With structural complexity, the metazoans have developed diverse adhesive and/or clinging organs, which have contributed to the existence of 62% ectoparasites. To get a comparative idea, the values reached for the different categories in metazoan and protozoan parasites are listed below:

Metazoa (see Pandian, 2021b)			
Ecto (%)	Gut (%)	Muscular (%)	Vascular (%)
62.2	32.3	5.0	0.5
Protozoa			
0	91.4	5.6	3.0

Strikingly, 91% protozoan parasites depend on the gut as their site of infection, whereas 62% metazoan parasites depend on ectoparasitism. Relatively, 6-times more protozoans are vascular parasites, in comparison to the metazoans.

TABLE 4.7

Number of marine protozoan parasitic species and their host species, as listed by some authors. *Adlard and O'Donoghue (1998)

Parasite species (no.)	Host species (no.)	Reference
Zoomastigophora		
Trypanosomatines: 50	Fish: 1000*	Lom (1984)
Sporozoa		
Gregarines: 33	Ascidians: 54	Monniot (1990)
Coccidia		
Aggregata group: 9 + 13* = 22	Cephalopods: 15 + 18 = 33	Hochberg (1990)
Non *Aggregata* group: 6 Haplosporea: 12 Microsporidea: 140	Chiton: 3 Bivalves: 11 Crustacea: ~ 15	Lauckner (1983) Lauckner (1983) Meyers (1990)
Ciliophora		
Ancistrocomids and sphenophyids 42	Bivalves: 33	Lauckner (1983)
16	Cephalopods: 39	Hochberg (1990)
19	Ascidians: 22	Monniot (1990)
1	Starfish: 4	Jangoux (1990)
Trichodina: 13	Bivalves: 12	Lauckner (1983)
354	1226	

A survey was also made to estimate the number of marine protozoan parasite species. As values assembled in Kinne (1983a, 1984, 1990) are scattered, the survey was limited to only those listed by some authors (Table 4.7). Notably, the list does not include values for the 78,000 speciose (see Pandian, 2017) marine gastropods host species. On the whole, some 1,226 marine host species are infected by 354 parasitic species. Notably, the value of 800 species estimated by Kinne (1983b) and the present one is far from complete. Yet, a few interesting reports are listed. 1. Of ~ 50 trypanosomatine species belonging to the genera *Trypanosoma* and *Trypanoplasmia* infect fish including the cod *Gadus morhua*. The largest *Trypanosoma gargantua* measures 130 µm. While their complete life cycle is not yet described, trypanosomatines are transmitted by leeches like *Calliobdella vivida*, *Johanssonia* sp. They are not pathogenic, as they occur in low density of 0.6 parasite/ml fish blood (Lom, 1984), in comparison to 5.2–5.5 × 10^8 pathogenic *Trypanosoma brucei brucei*/ml human blood (Harris et al., 1996). 2. Some *Hematodinium* spp are parasitic on pandalid shrimp and limit clutch size, while others are on crabs and amphipods, and cause 100% mortality in both sexes and all size classes of crabs. A climax seems to be that of *Thalassomyces californiensis*, a parasite on the Pacific shrimp *Pasiphaea emarginata*. The parasite penetrates the eyestalk and produces a root system that invades the brain and ventral nerve cord, and thereby kills the host shrimp (see Meyers, 1990). 3. Parasites are known to cause many defects including weight loss of the hosts. Interestingly, the infection by *Haplosporidium chitonis* increases body size of *Lepidochitona cinereus* (Fig. 4.1A). Incidentally, this seems to be common also for digenean parasites. For example, *Labratrema minimus* also induces increase in flesh yield in the cockle *Cardium edule* (Fig. 4.1B).

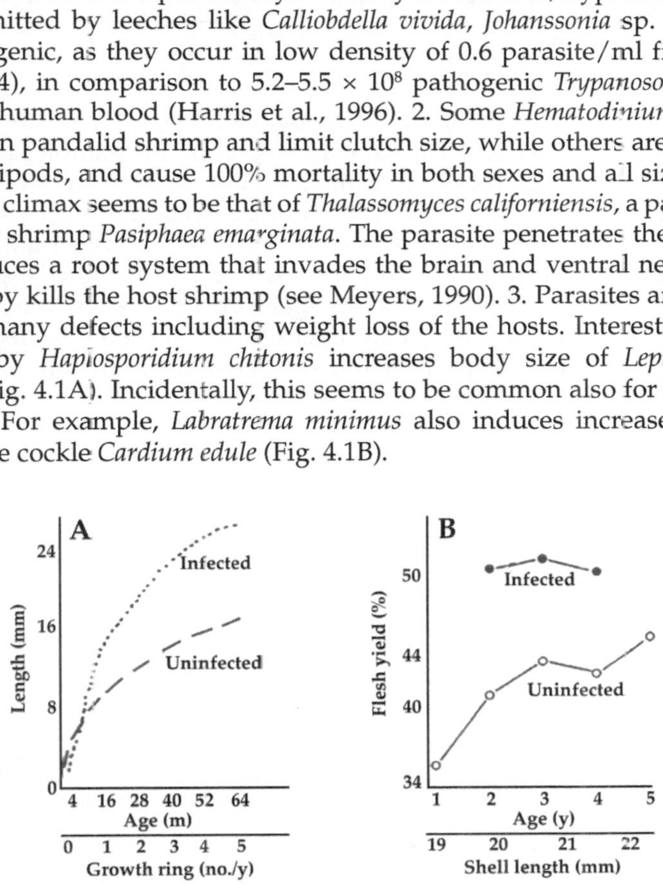

FIGURE 4.1

(A) Effect of *Haplosporidium chitonis* on shell growth of infected and uninfected *Lepidochitona cinereus* of different age and size classes (redrawn from Baxter and Jones, 1978). (B) Effect of *Labratrema minimus* on flesh yield of infected and uninfected cockle *Cardium edule* of different age and size classes (redrawn from Bowers, 1969).

TABLE 4.8

Species number of protozoan parasites on vertebrate host (modified [species number] from Adlard and O'Donoghue, 1998)

Class	Species (no.)	Incidence		Intensity	
		(no.)	(%)	(no.)	(%)
Pisces	32,500	1,000	3.1	2,400	2.4
Amphibia	5,228	1,500	28.7	200	0.1
Reptilia	9,545	1,300	13.6	600	0.5
Aves	10,038	4,000	39.8	700	0.2
Mammalia	5,513	2,000	36.8	2,800	1.4
Total/mean	62,824	9,800	15.6	6,700	0.7

The third survey by Adlard and O'Donoghue (1998) is limited to vertebrate hosts alone (Table 4.8). As their data are based on, for example, 4,000 avian species in 150 families collected from 63 countries, they are more reliable. Accordingly, the incidence of parasitic protozoan species is limited to 9,800 species or 15.6% of 62,824 speciose vertebrates. It increases from ~ 14% in slow motile reptiles to 39–40% in the fast-moving mammals and flying birds. On plotting the reported values against body size, as well as relative motility, it became apparent that intensity of infection decreases with decreasing body size except in mammals (Fig. 4.2). On the other hand, the

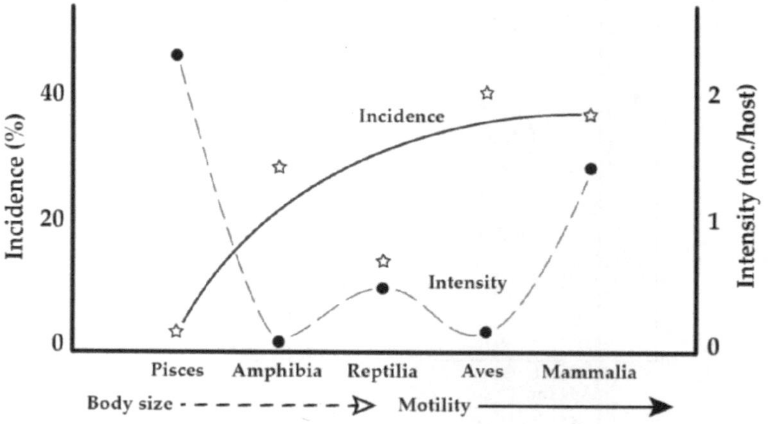

FIGURE 4.2

Incidence and intensity for protozoan parasitic infection in different vertebrate classes (based on Table 4.8).

incidence grows with increasing motility. Notably, birds with relatively smaller body size encounter one of the lowest (0.2 parasite/host) values for intensity of infection. For digenean parasites, more information is available on motility rate for both aquatic and terrestrial vertebrate hosts. In them, a similar analysis revealed that the incidence of digeneans decreases from ~ 800 helminths for the fast-swimming fish to > 50 helminths for slow-moving aquatic mammals (see Pandian, 2020). Hence, motility is the prime factor with regard to dispersal and incidence of parasites. The larger the body size, the greater may be the space to accommodate and provide resources for more numbers of parasites. Hence, intensity of infection depends on body size.

A comparative account was made on the relationship between commensal/symbiont/parasite on one hand and their respective hosts on the other. As one host may harbor more than 1 species of commensal/symbiont/parasite, the host-parasite ratio was considered more appropriate, keeping the host number as 1. The values indicate decreases in the following descending order: symbiosis > parasitism > commensalism. *Evolution seems to proceed more toward symbiosis rather than commensalism or parasitism.*

Source	Host: commensal/symbiont/parasite (species no.)	Host: commensal/ symbiont/parasite
	Commensalism	
Table 4.1	Bivalves: 9000* : Ciliates: 141	1: 0.016
	Symbiosis	
Table 4.2	Protozoa: 141 : Alga > 14	1 : 0.10
Table 4.3	Insects: 700 : Flagellates: 270	1 : 0.39
Table 4.4	Mammals: 180 : Ciliates: 500	1: 2.78
	Parasitism	
Table 4.6	Metazoa: 1.5 million : Protozoa: 32,950	1: 0.22
Table 4.8	Vertebrates: 62,824 : Protozoa: 9,800	1: 0.16

*Pandian (2017)

Transmission: For a parasite, transmission from one host to another is a critically risky event. Of 11,150 parasitic protozoan species (Table 4.6), the number of species involving two hosts in their life cycle may not exceed 1,800 or < 20% of parasitic protozoans, as the life cycle encounters twice riskier transmission. For want of relevant information, the quantification of parasitic species into the identified modes of transmission encounters the following problems: 1. For the 2,500 speciose ciliate parasitic species, information is available hardly for ~ 100 species. Many of them remain in a dual status as commensals and ectoparasites. (a) On contact with host the Pacific shrimp *Pasiphaea emarginata*, the parasitic ciliate *Thalassomyces californiensis* gains entry by penetrating the eyestalk. (b) Fig. 13.52 of Lauckner (1983) shows thigmotactic entry of the ciliate *Sphenophyra dosiniae* into the gill epithelial cell.

TABLE 4.9

Estimation on the number of parasitic/symbiotic protozoan species transmitted through different modes (based on Figs. 1.8B, 1.10B, 1.14–1.22, Lom, 1984, Hochberg, 1990, Monniot, 1990, Lauckner, 1983, Cali et al., 2017)

Taxonomic group		**Species**	
		(no.)	**(%)**
I. *Single hosted parasites*			
1. Direct from one host to other			
a) *Giardia vaginalis*, *Trichomonas* spp		55	
b) Ancistrocomids, Sphenophyids, *Trichodina* ciliates on bivalves		55	
c) Microsporidea		325	
	Subtotal	**435**	**4.7**
2. Trophic via sporozoite/cyst/spore			
a) Symbiotic/parasitic flagellates (- *Trypanosoma* spp)		1700	
b) *Entamoeba* spp, *Iodamoeba* spp		25	
c) Proteomyxidia		175	
d) Gregarinia (- *Nematopsis* spp)		1320	
e) Eucoccidia: Eimeriina		1445	
f) Haplosporea		51	
g) Myxosporidea		1100	
h) Microsporidea		650	
i) Symbiotic ciliates (Table 4.4)		488	
	Subtotal	**6954**	**75.7**
	Total for single hosted parasites	**7389**	**80.4**
II. *Double hosted parasites*			
1. Vector-borne: Injection			
a) Trypanosomatina	Fish/cattle/human – leeches/insects	100	
b) Haemosporina	Fish – leeches/crustaceans	156	
c) Piroplasmea	Equines – ticks	75	
	Subtotal	**331**	**3.6**
2. Vector-borne: Ingestion			
a) *Nematopsis* spp	Crabs – Oysters	330	
b) *Aggregata* spp	Cephalopods – Crabs	712	
c) Haemogregarinia	Fish – leeches/crustaceans	85	
d) Toxoplasmea	Felids – rodents/birds	1	
e) Microsporidea	Arthropods	325	
f) Ciliates: e.g. *Cromodina*	Euphausiids – squids	16	
	Subtotal	**1469**	**16.0**
	Total for two hosted parasites	**1800**	**19.6**
	Grand total	**9189**	**77.2**

(c) Another ciliate *Peniculistoma mytili* enters the foot of *Mytilus edulis* by thigmotactic cilia. Hence, ciliates are considered to gain transmission by direct contact. 2. Myxosporideans are reported to gain transmission through spores. On rupture of the infected cell of the gut or its associated organ (e.g. gall bladder) or renal organ, the spores are released to the exterior. On being ingested, the germ or sporoplasm is released by eversion of the polar filaments. However, it is not clear how the spores of muscle-encysting species like *Kudoa* are released to the exterior. The spores may be released either on the death of infected host or on ingestion of the infected host by a benign host. Hence, all the 1,100 myxosporidean species are considered under the group trophic transmission via spores (Table 4.9). 3. Regarding the 1,300 speciose Microsporidea, they are grouped into the following categories. Some of them gain entry on contact and penetration (e.g. *Nucleospora, Enterospora*), some others are trophically transmitted through spores and yet others are two-hosted parasites (e.g. *Amblyospora*, Fig. 1.19). For want of relevant information, some 325 species are assigned to each of the Group 1 Direct contact and Group with two hosted parasites. The remaining 650 species are considered under the group, in which trophic transmission occurs via the spores.

In all, the modes of transmission could be identified for 9189 species only or 77.2% of the 11,901 protozoan species (i.e. 11,150 parasitic + 751 symbiotic and commensal species). Understandably, *7,078 species or 64% of them involve one host, while 16% of them involve two hosts* (Table 4.9). Within the former, *trophic transmission mode through sporozoite* (e.g. *Lecudina tuzetae*), cyst (e.g. *Nematopsis ostrearum*) or spore (e.g. *Amblyospora*) *has a lion share of 75.7%. Among the two hosted parasites, the Definitive Host (DH, ~ 1,464 species, 16.0%) are infected, when they actively ingest the infected Intermediate Host (IH).* Transmission takes place in 331 species or 3.6% protozoan parasites through sanguivorous flies, mosquitoes or ticks. The transmission via direct contact by parasitic species gaining entry through penetration may be a costlier mode.

5

Sexual Reproduction

Introduction

1. Sex is a luxury and costlier. For example, a completely clonal Protozoa may divide 5–100 times faster than its sexual counterpart. On this basis, Lewis (1983) estimated the cost of sexual reproduction as 5 to 100-fold costlier. 2. However, segregation during meiosis and recombination at fertilization provides a greater scope for generation of new gene combinations and thereby accelerates the processes of evolution and speciation. Nevertheless, meiosis demands the establishment and maintenance of cellular mechanism and usually requires a longer duration than mitosis. 3. In the absence of physical separation between germ and soma, Protozoa can ill-afford true metazoan-like meiosis. In fact, males contribute their genetic material alone to the offspring. In metazoans, this is why evolution seems to have used them in parental care. 4. Metazoans can also skew female-biased ratio, when more progenies can thrive under favorable condition, or male-biased ratio, when their females need to select more competent males for mating to produce more fit progenies. 5. Protozoans can ill-afford the resource-demanding showy structure to attract mates (e.g. like in metazoans). They only depend on chemical cues to attract sperms (see El-Bawab, 2020). However, a mating dance is performed by pre-conjugants of the hypotrich ciliate *Stylonychia* (see Grell, 1973). The unicellularity does not allow the adoption of some of these strategies (Stelzer, 2015). It is in this context, that protozoan sexual reproduction may have to be considered.

All organisms are subjected to random mutations, which generate new gene combinations. In metazoans, true sexual reproduction involves a generation of new gene combinations during (meiotic) gametogenesis and recombination at fertilization. As a result, they gain new gene combinations both through random mutations and sexual reproduction. Plants provide excellent examples for the rate of speciation in sexless cyanobionts, and sexualized algae and bryophytes. The number of years required to 'create' a species is 0.758 million years (MYs) for the 3,300 speciose Cyanophyceae, which depend on random mutations alone for generation of new gene combinations. This value is 0.24 MYs for the 32,777 speciose Chlorophyceae, in which the 'vegetative' haploid gametes are mitotically generated. In them, there is less scope for generation of new gene combinations as mitosis is followed by fertilization. The more advanced 21,925 speciose sexualized bryophytes require only 0.2 MYs to generate a new species. In them too, the motile flagellated gametes are generated through mitosis and fertilization precedes meiosis. With inclusion of meiotic gametogenesis, only 390 years are required to generate a species in angiosperms (see Pandian, 2022). Clearly, *the acquisition of sexual reproduction accelerates the processes of evolution and speciation*. Some protozoans simply transform the 'vegetative' cells into 'gametes' (e.g. the hologamic cryptomonadids, coccolithophores [Fig. 1.7], arcellinids [Fig. 1.9A] and conjugating ciliates [Fig. 1.23]). In the latter, the micronuclei act as 'gametes'. In the autogamic *Actinophrys* (Fig. 1.9B), a different pattern of meiosis occurs, albeit the haploid gametes of the same parent are fused.

In Protozoa, there are several routes, through which the sexual life cycles are completed. At least, 30 cycles are illustrated in Fig. 1.7 to Fig. 1.26. A look at them may provide an idea about the trials and tribulations undergone by protozoans during the checkered history of evolution. In fact, none of them seems to have ever discovered the typical metazoan-like pattern of gametogenesis, in which diploid gametocytes undergo a highly conserved route of meiosis to generate four viable sperms and one fertilizable ovum from a spermatocyte and an oocyte, respectively. Considering these features, 'sex' and 'sexuality' of protozoans remain a debatable subject (see Stephen, 1990). Though sexual reproduction involves gametogenesis, this chapters is devoted to 'sexual' reproduction in Protozoa in the context of species diversity. Chapter 7 elaborates gametogenesis in them.

5.1 Sex and Sexuality

In an individual, sexualization is recognized by the complimentary occurrence of meiosis and fertilization (Hawes, 1963). 1. As protozoans do not exhibit the typical metazoan-like meiosis, this account considers category I as consisting of these non-sexualized protozoans, which use 'vegetative cells as gametes', category II comprises sexualized but agametogenic protozoans, and category III includes those protozoans, which have secondarily lost sex. On losing the flagella and chloroplasts (see Kuroiwa et al., 1993), the vegetative dinoflagellates are transformed into 'gametes'. *Some flagellates (2,300 species), rhizopods (2,281), haplosporeans (51) and protociliates (200) remain non-sexualized* (Table 5.1).

TABLE 5.1

Estimation on the number of species belonging to different categories of sexuality (species numbers alone are drawn from *onezoom.org*)

Taxa	Species		Taxa	Species	
	(no.)	(%)		(no.)	(%)
I. Non-sexualized Protozoa					
1. Mastigophora			2. Rhizopoda		
a. Chryptomonadids	100		a. Amoebida	181	
b. Chrysomonadids	100		b. Arcellinida	2000	
c. Coccolithophora	< 100		c. Heliozoa	100	
d. Dinoflagellida	2000		Subtotal	2281	
Subtotal	2300		4. Ciliophora		
3. Sporozoa			Protociliates	200	
Haplosporea	51		**Total**	**4832**	**14.7**
II. Sexualized agametogenic Protozoa					
1. Chloromonadida	500		**Subtotal**	**8300**	**25.2**
2. Euciliata	7800				
III. Secondarily sex lost Protozoa, (see also Doerder, 2014)*					
1. Mastigophora			3. Sporozoa		
a. Euglenida	1000		a. *Nematopsis* spp	330	
b. Parasitic flagellates	1900		4. Ciliophora		
Subtotal	2900		a. *Tetrahymena**	47	
2. Rhizopoda			**Total**	**3296**	**10.0**
a. Amoebae	19		**Grand total**	**16428**	**49.9**
			Sexualized protozoa (32,950–16428):	**16522**	**50.1**

The ciliates are a peculiar taxonomic group. They have no mechanism for gametogenesis but are monoeciously or dioeciously sexualized, as they can generate through meiosis motile and immotile micronuclei, which act as gametes to perform oogamous fusion. Hence, they are considered as sexualized agametic group (Table 5.2).

Hawes (1963) was the first to recognize that a few flagellates, amoebae and the ciliate *Tetrahymena* rely only on clonal multiplication. They are considered to have secondarily lost sex. He traced the reason for the loss of sex to polyploidy. For example, *Amoeba proteus* has 500–600 chromosomes. He noted that the loss of sex is rare among Ciliata and unknown in Sporozoa. However, the loss of sex also does occur in sporozoans. For the first time, through a survey, this account has identified their incidence in all the major four classes of Protozoa and quantified their species number. Accordingly, 3,296 species or 10.0% protozoans have secondarily lost sex; mostly mastigophores and sporozoans incurred the loss (Table 5.1). On the whole, 16,428 species or 49.9% protozoans remain either non-sexualized or sexualized agametogenics or anamorphics, i.e. secondarily lost sex.

In protozoa, sexuality includes monoecy (hermaphroditism) or dioecy (gonochorism). Considering the presence of isogamic or anisogamic gametes, monoecy or dioecy is recognized. From cultivation study of K.G. Grell, Tubingen, Hawes (1963) noted that of 8, 4 foraminifer species (e.g. *Glabratella sulcata*) are cross-breeders and *Rubratella intermedia* is an anisogamic. Hence, 50% of the 4,500 speciose foraminifers are considered as dioecious. Regarding the 1,100 speciose Myxosporidea, Hyman (1940) hinted at the presence of isogamy and anisogamy. Figure 1.22 indicates that 50% of them are isogamics. From Figures (Fig. 1.7 to 1.26) illustrating the life cycles, the isogamic monoecy or anisogamic dioecy is recognized. For the 125 speciose choanoflagellates, Fig. 3 of Levin and King (2013) clearly shows that they produce anisogamic gametes. Therefore, they are considered as dioecious. By culturing 18 volvocid species, Smith (1944) found 11 and 7 species as dioecious and monoecious, respectively. On the basis of Smith's report, the values 156 and 244 species were recognized as homothallic monoecious and heterothallic dioecious, respectively. Chloromonadida consists of 500 species, in which some are monoecious, while the others are dioecious (El-Bawab, 2020). Considering the description in p 99, 250 species or 50%

TABLE 5.2

Estimations on the number of monoecious and dioecious protozoan species (species number drawn from *onezoom.org*)

Taxa	Species		Taxa	Species	
	(no.)	(%)		(no.)	(%)
I. Monoecy			II. Dioecy		
a. Volvocids	156		a. Volvocids	244	
b. Filosians	300		b. Choanoflagellates	125	
c. Foraminifers	2250		c. Foraminifers	2250	
d. Radiolarians	4200		d. Other sporozoans	4819	
e. Parasitic rhizopods	250		**Total**	**7438**	**49.1†**
f. Myxosporideans	550		**Grand total**	**15142**	**46.0**
Total	**7706**	**50.9†**	**†as % of 15,142 species**		

chloromonadids are considered as monoecious (e.g. *Chlamydomonas snowiae*, *C. reinhardtii*, *C. moewusii*), while the others (e.g. *C. braunii*, *C. coccifera*, *C. ooganum*) are considered as dioecious. However, they are not gametogenics.

On the whole, of 15,142 species, 49.1% protozoans are dioecious, whereas 7,706 species or 50.9% protozoans are monoecious. Incidentally, this survey recognizes different types of sexuality in 16,428 + 15,142 = 31,570 species or 96% protozoans.

5.2 Nuclear Fusion and Fertilization

Considering the various reproductive cycles in protozoans, Anderson (1983) grouped them into (i) gametogamy, (ii) autogamy and (iii) gamentogamy. The ensuing description will show that this grouping is porous, as autogamy occurs within hologamy (e.g. Heliozoa) and gametogamy within hologamy (e.g. *Chlamydomonas*). Before the description, a few terms may be explained. 1a. Hologamy involves the association of 'vegetative individuals' acting as 'gametes'. The association commences with attachment of two eligible individuals. It is followed by changes in nuclear morphology (see Fig. 1.9B). Many authors including Hyman (1940) considered that the nuclei of the partners fuse to form a zygote (e.g. cryptomonads, coccolithophores, Fig. 1.7, arcellinids, Fig. 1.9A). However, cytological or molecular evidence for the fusion is not available. Notably, out of two partners, only two progenies arise. 1b. In the hologamic dinoflagellates, the eligible two (haploid) partners come together and establish a conjugation tube to facilitate the fusion of the two nuclei and chloroplasts of the conjugants to form a planozygote (Salmaso and Tolotti, 2009). Here also, only two offspring appear from two conjugants. In some dinoflagellates, the encysted planozygote undergoes meiosis to generate four offspring, in each of which mitosis is immediately followed to produce eight offspring out of two conjugants.

1c. What may be called as hologamic autogamy occurs in Heliozoa. It involves two successive meioses to generate two viable 'gametes' and six degenerated polar bodies. Followed by isogamic fusion of the two gametes, diploid zygote is formed – all within a cyst. Eventually, only two diploid progenies emerge from the cyst through binary fission – as in cryptomonadids. However, the scope for generation of new gene combination is limited, as the two offspring appear from the same single parent. Also in ciliates, the two conjugating partners produce only two progenies (e.g. *Paramecium aurelia*) or four progenies (*P. caudatum*). *Being costlier, sexual reproduction in Protozoa can afford to produce only fewer progenies.* Hologamy may also involve gametogamy, as in Chloromonadida. In them, the biflagellated motile zoospores are produced by the simple transformation of the 'vegetative' individuals. It is now known that M+ and M– mating types are approximately equal to female and male gametes, respectively (see Pandian, 2022).

Flagellated motile haploid isogamic gametes are generated by foraminifers and radiolarians; in the former, the gamete number is limited to eight only but numerous in the latter. In them, fertilization may occur within the gametes generated by a single parent or between two parents. Except for 330 speciose gregarines namely the *Nematopsis* group (Fig. 1.16A), all other sporozoans do undergo gametogamy. *Barring the 51 speciose Haplosporea and 1,300 speciose Microsporidea, all other sporozoans are anisogamic gametogamics.*

5.3 *Chlamydomonas* – Volvocids

In the 150 speciose genus *Chlamydomonas* (*Wikipedia*) and 400 speciose order Volvocida, gametogenesis and fertilization simulate those of metazoans. However, they generate gametes through mitosis and not by meiosis. Further, the gametes in the former appear by a simple transformation of 'vegetative' individuals. Interestingly, the chlamydomonads and volvocids present a contrasting picture. In the former, the biflagellated motile gamete types range from isogamy (e.g. *Chlamydomonas snowiae, C. reinhardtii, C. moewusii*) to anisogamy (e.g. *C. braunii*, Fig. 5.1A) and to oogamy (e.g. *C. coccifera, C. ooganum*). In *C. braunii*, one partner produces eight micro mating type, while the second parent generate four macrogametes. Both mating types are flagellated and motile but the micro mating type is more active. In all of them, two mating types or zoospores are produced. These mating types can be identified by the presence of two or three electron dense layers, which lie under the plasma membrane in M– and M+ zoospore, respectively (see El-Bawab, 2020). Both these mating types can be produced from the same homothallic or preferably monoecious mother, as in *C. snowiae, C. reinhardtii*, or from different heterothallics or dioecious parents, as in *C. braunii, C. oogamum*. Their pairing is always between M+ and M– mating types. On clumping, the flagella of the two mating types twist around each other, enabling specific region of the plasma membrane to come together and agglutinate (Fig. 5.1B). Following the flagellar agglutination, the M+ gamete initiates fusion by extending a fertilization tube, through which the fusion occurs. The mating types may fuse face-to-face (e.g. *C. moewusii*) or shoulder-to-shoulder (e.g. *C. reinhardtii*). After the fusion, the quadri-flagellated diploid zygote may also remain motile for several hours and in some like *C. pertusa* up to 15 days prior to becoming an encysted zygospore. Under favorable conditions, the zygospores meiotically divides to produce four haploid zoospores, which eventually develop into mature vegetative individuals (see El-Bawab, 2020).

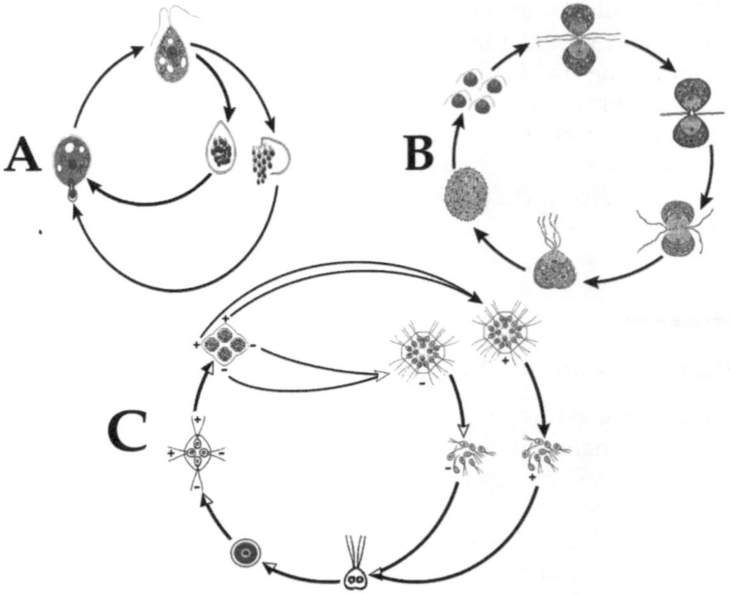

FIGURE 5.1

Life cycles: (A) Anisogamic oogamous *Chlamydomonas braunii*, (B) isogamic mating in *C. snowiae* (redrawn from Pandey and Trivedi, 1995) and (C) the dioecious colonial flagellate *Gonium pectorale* (modified from Grell, 1973).

The multicellularity has facilitated the formation of biflagellate motile sperm in antheridium and immotile egg in oogonium at adjacent locations sunken from the outer spherical layer of the colony. The formation of gametes and the fertilization was described earlier (Fig. 1.8A). Figure 5.1C illustrates the life cycle of the 16-celled *Gonium pectorale*. In this dioecious colony, the two M+ and M– mating types are generated from two parental colonies. According to Grell (1973), the diploid zygote meiotically generates the 4-celled colonial progenitor, in which two of them are M+ type and the other two are M– type. Subsequent mitoses produce 16-celled with either M+ or M– mating type. However, it is also possible that the two nuclei from the mating type M+ and M– may remain separately intact within a zygote and mitotically produce their own respective mating types.

6

Clonal Multiplication—Regeneration

Introduction

Sex is a luxury and costs time and energy but ensures recombination to generate new gene combinations (Carvalho, 2003). As benefits arising from the recombination outweigh the cost of time and energy, sex was evolved as early as 1.6–2.0 billion years ago (BYA) (Butlin, 2002) and is manifested in a wide range of microbes, plants and metazoans. Notably, cyanobionts, which arose 2.5 BYA, remain sexless and multiply clonally alone. In fact, all microbial organisms inclusive of cyanobionts, which originated prior to 1.6–2.0 BYA, remain sexless. It is likely that the non-sexualized 4,832 protozoan species spread over the flagellates, amoebids and ciliates (Table 5.1) appeared before the discovery and manifestation of sex ~ 2.0 BYA. For them, clonal multiplication is the most common mode of reproduction. However, the costlier sexual reproduction also rendered easier access to the invasion by clonality in many plants and animals, especially protozoans. As a result, (i) clonality is an obligate equal component in heterogonic life cycle in foraminifers. (ii) In other protozoans, it plays a dominant role. For example, clonal mode of reproduction in (a) the amoeba *Entosiphon sulcatum* (see Hyman, 1940) and (b) the ciliate *Paramecium aurelia* (Woodruff, 1926) can go on over 947 and 1,500 generations, respectively, uninterrupted by sexual reproduction. (iii) Strikingly, ~ 3,296 species or 10.0% protozoans have secondarily lost sex; they multiply clonally alone (Table 5.1).

6.1 Binary Fission

Most protozoans clonally multiply through binary fission. Before the fission or division, they, for example, the flagellates withdraw the flagellum, slightly shrink the cytoplasm within the wall and incur the disappearance of contractile vacuole (see El-Bawab, 2020). In the oval-shaped flagellates like *Euglena spirogyra*, the symmetrogenic division involves the flagella, cytopharynx, chloroplast and nucleus and produces the mirror-imaged clones of two daughter progenies (Fig. 6.1A–B). When the division is rapid, as in some chlamydomonads, one of the daughters may not inherit the chloroplast, due to its slow and incomplete division. Unable to regenerate it, the chloroplast-less daughter may transmute from *Chlamydomonas* to *Polytoma* (Borradaile et al., 1977). (i) Within the flagellates, the dinoflagellates undergo asymmetrogenic transverse division, cause unequal inheritance of maternal traits and generate two unequal daughters. However, the daughter has the potency to regenerate the missing maternal components (Fig. 6.1C–E). (ii) In arcellinids (e.g. *Difflugia*, Fig. 1.9A), one daughter gains the parental shell but the other has to build a new one. (iii) In ciliates, which mostly divide transversely, the clonal daughters inherit differently with regard to traits like the cytostome apparatus, cytopyge, stalk in sessiles and so on. Still, all these clonal daughters can regenerate the missing maternal components. In clonal multiplication involving binary fission, the traits are usually equally inherited by the offspring in many protozoans; the traits are, however, differently inherited by the offspring in others. (iv) Added to these, the cytoplasmic inheritance of killer trait in *Paramecium* (Sonneborn, 1947) have all raised an apprehension in distinguishing sexual from clonal reproduction. Further, protozoans are deemed to be immortal (see Stephen, 1990), *as their unicellularity does not facilitate the division between the 'soma' and 'germ' components*. In metazoans, sexuality inherently results in mortality, while protozoans may be immortal, albeit undergoing senescence.

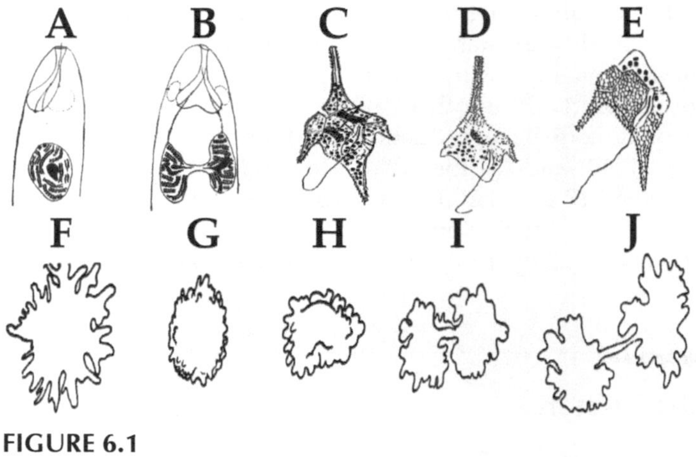

FIGURE 6.1

(A–B) Longitudinal division in *Euglena spirogyra* (after Ratcliffe, 1927), (C–E) transverse fission in *Ceratium* (after Jorgenson, 1911), (F–J) different stages of binary fission in *Amoeba dubia* (after Botsford, 1926).

Irrespective of longitudinal or transverse division, the binary fission generates two progenies out of a parent, as in euglenids (e.g. *E. spirogyra*), dinoflagellates (e.g. *Ceratium*), amoebids (e.g. *Amoeba dubia*, Fig. 6.1F–J), arcellinids (e.g. *Difflugia*, Fig. 1.9A) and ciliates (e.g. *Paramecium*, Fig. 1.23) but by following consecutive binary fissions, four in radiolarians (e.g. *Anlacantha scolymantha*, Fig. 1.13), 12 schizonts in foraminifers (e.g. *Elphidium crispum*, Fig. 1.12). In *Chlamydomonas*, the parent divides consecutively twice but rarely thrice to produce two, four or eight progenies from a parent (see El-Bawab, 2020). It is common among flagellates that following repeated multiplications, some daughters lose flagella, and form a gelatinous aggregation, known as palmella stage.

6.2 Multiple Fission

In Protozoa, the division rate depends on (i) nutrient availability, (ii) body size and (iii) temperature. The assured nutrient availability and size reduction by successive divisions facilitate multiple fissions. With storage of nutrients, the encystation renders multiple divisions. For example, it is followed by multiple divisions resulting in the emergence of ~ 32 progenies in *Amoeba proteus* (Fig. 1.10A). Emerging from the cyst, the planozygote of some dinoflagellates undergo 'meiosis' to produce four daughters, each of which immediately mitotically divides to produce eight progenies. The assured nutrient availability allows multiple divisions in parasitic protozoans. For example, the micronucleus of the holotrich ciliate *Ichthyophthirius multifiliis*, a skin parasite of fish, divides as many as 1,000-times (see Hyman, 1940). In parasitic sporozoans, repeated multiplications produce many infective sporozoites. Borradaile et al. (1977) named this division process as sporogony and its products as spores. The ensuing list shows that *evolution has proceeded toward increasing the number of sporozoite/cyst with the shift from gut to corpuscular parasitism*. The increase is from 2–12 sporozoite/cyst in gregarines to 3,700–10,000 sporozoite/oocyst in Haemosporina.

Species name	**Sporozite/cyst or oocyst (no.)**	**Species name**	**Sporozite/cyst or oocyst (no.)**
Gut parasites			
Nematopsis ostrearum	2 (Fig. 1.16A)	*Eimeria*	4 (Fig. 1.17)
Schizocystis	8 (Fig. 1.15B)	*Haplosporidium*	32 (Fig. 1.20)
Lecudina tuzetae	12 (Fig. 1.15A)	*Toxoplasma gondii*	32 (Fig. 1.21)
Aggregata spp	3–28 (see p 35)	*Amblyospora*	4 (Fig. 1.19)
Corpuscular parasites (see Simonetti, 1996)			
Plasmodium vivax	3,700	*P. falciparum*	10,000

Incidentally, the regenerative potency of protozoans may be described. 1. The natural longitudinal or transverse as well as binary or multiple divisions, produce two or more number of daughter progenies in all the clonal flagellates, rhizopods and ciliates; each of their offspring is completely regenerated. This clearly indicates that these protozoans have the potency for complete regeneration. 2. Parasitic protozoans seem to have lost the potency. The rapidity (cf transmutation of *Chlamydomonas* to *Polytoma*, see p 102), at which daughter progenies or spores are generated through sporogenesis can be a reason for the loss of their regenerative potency by sporozoans. 3. In cut pieces of rhizopods and ciliates, complete regeneration is achieved in all nucleated pieces with sufficient cytoplasm (see also Balamuth, 1940). 4. Hyman (1940) narrated a few interesting experimental observations reported by early workers like Sokoloff (1924). 4a. Isolated nuclei are unable to generate cytoplasm and perish (e.g. *Amoeba proteus, Gromia fluviatilis*, see Balamuth, 1940). Arguably, this indicates the nuclear inability to 'create' cytoplasmic mRNA system required for protein synthesis. 4b. In enucleated fragments of *Uroleptus* and *Uronychia*, regeneration is initiated but not completed, indicating the obligate need for nucleus to complete regeneration. Therefore, both nucleus and a fraction of cytoplasm are obligately required to complete regeneration. For details on flagellar regeneration, Rosenbaum and Child (1967) may be consulted. 4c. In the ciliates *Stentor, Spirostomum, Dileptus, Uroleptus, Euplotes* and others, a cut piece of one fiftieth size or even a piece as small as one seventy fifth can complete regeneration, so long each of these pieces contains nucleus and adequate quantum of ectoplasm to cover the cut surface. The publication by Aufderheide et al. (1980) seems to be the last contribution on this subject.

6.3 Solitary Fragmenters vs Colonial Budders

For the first time, a survey was made to estimate the number of metazoan clonal species that are solitaries or colonials as well as fragmenters or budders. Surprisingly, of 32,519 metazoan clonals, as many as 29,464 species or 95.2% are colonials but only 1,490 species or 4.8% are solitaries. More surprisingly, almost all the fragmenters are solitaries, while budders are all colonials. Further, fragmentation is a costlier mode of clonal multiplication than budding. Consequently, fragmentation decelerates species diversity but budding accelerates the diversity (Pandian, 2021b). An attempt has been made to extend these findings for metazoans to protozoans. The parasites Mastigophora (1,900 species), Rhizopoda (250 species), Sporozoa (6,500 species) and Ciliophora (2,500 species, see Table 4.6) are not considered in this survey. However, the grouping of the other protozoans into solitary fragmenters and colonial budders is refuted by the fact that some solitary sessiles like *Ephelota gigantea, E. gemmifera* are budders and the branched motiles like *Dendrocometes* are also budders. That necessitated at least genus level search for the sessile suctorians and peritrichids.

Before the search, the budding process, in suctorians may be described. Attached to the substrate by a non-contractile stalk, the suctorians ingest food through tentacles and reproduce through the formation of buds, which are eventually released as ciliated swarmers. Budding commences with the invagination of a small pericellular area with barren basal bodies. Some of these bodies are incorporated into the swarmer-anlage and give rise to eight rows of cilia (Bardele, 1970). Anderson (1988) recognized three types of budders: (i) surface budding (*Ephelota gigantea*, Fig. 6.2A), (ii) budding initiated within a pouch but completed at the surface (*Actineta tuberosa*, Fig. 6.2B) and (iii) budding completed within the pouch (*Dendrocometes*, Fig. 6.2C). Suctorians settle on motile hosts; for example, *Ephelota plana* infest the krill *Euphausia pacifica* and differentiate into 0.54 ♀ : 0.46 ♂. Most often they are harmless but the lethal *Pseudocollinia* spp kill the euphausiid host within 48 hours of infection (see Endo et al., 2017).

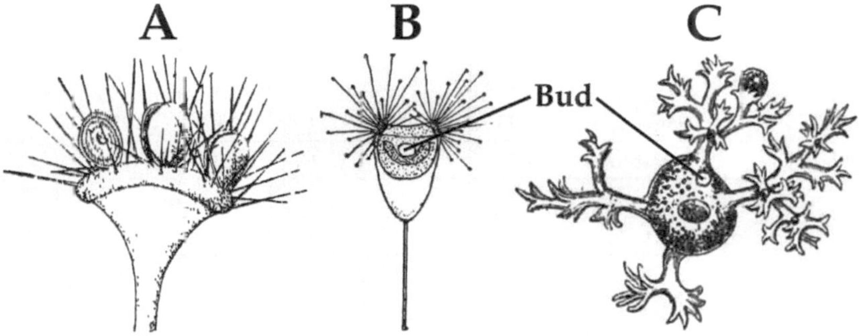

FIGURE 6.2

Ephelota gigantea, (B) *Actineta tuberosa* and (C) *Dendrocometes* (after Noble, 1929, after Kent, 1881, Pestel, 1931).

Of 1,021 peritrichids, 683 species are colonials; the others (338 species) are solitaries (Table 1.12). They clonally multiply by longitudinal fragmentation, as the solitary choanoflagellates do it (Fig. 6.3B). Figure 6.3A illustrates clonal multiplication through budding, development of juveniles, and hatching in a floating colonial volvocid. Of 125 sessile choanoflagellates, there are 108 solitary and 17 colonial species, respectively. In colonials, individual flagellates isolate into 'swarmers' to ensure dispersal; on settling, they clonally multiply to form the colony (Fig. 6.3C). Radiolarians clonally multiply by binary fission (see Fig. 1.13).

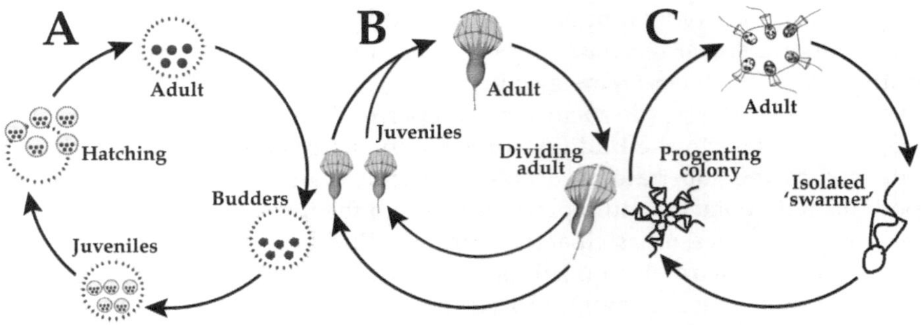

FIGURE 6.3

Clonal multiplication by: (A) budding in a colonial volvocid (based on Umen, 2020), (B) longitudinal division in a solitary choanoflagellate and (C) through isolated 'swarmer' in a colonial choanoflagellate (based on Larson et al., 2020).

The values reported for sessiles in Table 1.12 are summarized hereunder:

Taxa	Solitary		Colonial		Total
	(no.)	(%)	(no.)	(%)	
Mastigophora					
Volvocids	341	87*	52	13*	393*
Choanoflagellates	108	86**	17	14**	125**
Subtotal	**449**	**87**	**69**	**13**	**518**
Rhizopoda					
Radiolarians	–		244		
Ciliophora					
Peritrichida	470	33†	950	67†	1420†
Suctoria	152	32‡	328	68††	480††
Subtotal	**622**	**33‡**	**1278**	**67‡**	**1900‡**
Grand total	**1071**	**41**	**1531**	**59**	**2602**

* as % of 393 species; ** as % of 125 species, † as % of 1420 species; †† as % of 480 species; ‡ as % of 1900 species

The values assembled for sessile ciliates alone are underestimates. For example, the assembled value is 347 species or 72% of the 480 speciose Suctoria (*onezoom.org*). Hence, the values for ciliates alone are revised as 1,420 and 480 species for the Peritrichida and Suctoria, respectively. From the values summarized, the following may be inferred. 1. *Of 32,950 protozoans, 1,494 species or 4.5% protozoans are sessiles*. Among metazoans, 44,580 species or 2.9% of metazoans are sessiles. 2. *Of 1,368 sessile ciliates, 452 species or 33%*

are solitaries, whereas 67% are colonials. In 32,519 metazoan clonal species, 4.8% are solitaries, while 95.2% are colonials. 3. Remarkably, 87% mastigophores are solitaries, whereas > 67% ciliates are colonials. Hence, ciliates are better adapted to coloniality; for example, with contractile stalk in peritrichids, individuals in a colony can feed at different levels in the water column.

4. Budding is limited to 52 species in the floating motile volvocids. But there are differences between floating colonial volvocids and sessile ciliates. Arguably, all the sessile ciliates – irrespective of their solitary or colonial status – may have to go through a heterogonic life cycle, in which clonal multiplication is obligately alternated with sexual reproduction. Accordingly, the buds are differentiated into the dispersive motile ciliated swarmers, which are terminated as conjugants leading to sexual reproduction and eventual settlement (see Fig. 1.22B). Therefore, *it is difficult to distinguish sessile ciliates into solitary fragmenters and colonial budders*, as it is possible among the metazoan sessiles. Further, there are also succinct differences between peritrichids and suctorians. Firstly, the solitary peritrichids like *Vorticella* (Fig. 1.22A) are capable of clonal multiplication by vertical fragmentation as well as budding. A cursory survey over the images of solitary peritrichids (e.g. *Cothurnia, Epistylis, Platycola, Thuricola, Stentor, Vaginicola*) revealed that almost all the *solitary peritrichids are capable of clonal multiplication by fragmentation and budding. Therefore, the scope for numerical diversity may be far greater for solitary peritrichids than that for those of suctorians, which can clonally multiply only by budding*. Secondly, in 50% of the solitary peritrichids (e.g. *Carchesium, Vorticella, Zoothamnium*, Randall and Hopkins, 1962), *the contractile stalk enables filter-feeding at different levels in the water column as well as escape from predators. These two features of the peritrichids seem to have accelerated the diversity to 1,027 species. Their absence may have decelerated the diversity to 347 species in Suctoria.*

7

Gametogenesis

Introduction

Gametogenesis is a process, through which gametes are generated. Being the parent cells in metazoans, the diploid oogonium and spermatogonium can alone undergo meiotic division to generate one haploid oocyte and four haploid spermatozoa, respectively. The process involves segregation of parental chromosomes and their traits. It provides the basis for recombination at fertilization. Thereby, it generates new gene combinations – the raw material for evolution and speciation. In Protozoa, it differs widely from one taxonomic group to the other, indicating the trials and tribulations undergone by them. Yet, the unicellularity has not let any one of them to establish the metazoan-like meiotic gametogenesis. The categorization of protozoan gametogenics may appear a repetition of the chapter on Sexuality. But it is necessary and includes groups within category I. In them, the pictorial representation of gametogenesis may also appear to simulate the vertical representation of their life cycles. But they are useful to get a comparative picture.

7.1 Mitosis – Meiosis

Within Protozoa, this account recognizes I. Vegetative gametic Protozoa, that do not have the mechanism for gametogenesis and employ the transformed vegetative cells as gametes, II. Agametic sexualized Protozoa like the ciliates and III. Secondarily sex-lost Protozoa like the euglenids and IV. Sexualized gametic Protozoa, which have the mechanism for gametogenesis. Within the category I, three groups are recognized: Group 1 consists of Protozoa that produce vegetative gametes through mitosis alone (e.g. most dinoflagellates). In Group 2, mitosis is consecutively (e.g. coccolithophores) or subsequently (e.g. haplosporeans) followed by meiosis. Group 3 comprises protozoans, in which meiosis is commenced to produce vegetative gametes and it is followed by mitosis, as in some amoebids.

The Ciliophora are a peculiar taxonomic group. They have no mechanism for gametogenesis, as in sexualized protozoans. Within the conjugants, motile and immotile micronuclei are generated by meiosis. Their micronuclei act as gametes to achieve oogamous fusion. As it is difficult to place them as vegetative gametic protozoans or sexualized gametic protozoans, they are separately considered under item II as sexualized agametic Protozoa.

Hawes (1963) recognized that in protozoans, sexualization can be identified by the complimentary occurrence of meiosis and fertilization. The fact that the non-typical 'meiosis' do occur in the life cycle of protozoans, which are characterized by vegetative gamete production indicates that the idea of Hawes (1963) on sexualization may not be correct.

Table 7.1 shows that vegetative gametogenesis is scattered among the recognized three major categories as well as in groups within the recognized categories, suggesting that Protozoa are polyphyletics and are assembled together for taxonomic convenience. Within the 4,100 speciose Category I, Group 1 includes mostly dinoflagellates, arcellinids and others (Table 7.1). Group 2, in which vegetative gametes are produced by mitosis followed by meiosis; in it, there are 332 species only. Group 3 comprises only 400 species. *On the whole, 4832 species or 14.7% protozoans have no mechanism for gametogenesis and their transformed vegetative cells act as gametes.*

Being a peculiar group, 7,800 ciliate species or 23.7% of protozoans are brought under category II, in which sexualized protozoans are not gametogenic. To this category, the 500 speciose Chloromonadida may have to be added, as they are also sexualized (Fig. 5.1A) but many of them are not gametogenics.

Surprisingly, some protozoans are reported to have lost sex secondarily. Hawes (1963) was the first to recognize it, especially in a few flagellates, amoebids and ciliates. Through the first survey, this account has identified and quantified the taxa that have secondarily lost sex and the mechanisms for gametogenesis, as well. Table 7.1 lists their incidence in euglenoids, parasitic flagellates, coccidians. *As much as 10.0% or 3,296 protozoan species have secondarily lost sex and mechanism for gametogenesis.*

TABLE 7.1

Estimation on the number of species belonging to different categories of gametogenics (values from *onezoom.org*). * see Table 5.2

Taxa	Species		Taxa	Species	
	(no.)	(%)		(no.)	(%)
I. Vegetative gametogenic Protozoa					
Group 1: Mitosis only			Group 2: Mitosis + Meiosis		
a. Chryptomonadids	100		e. Coccolithophores	< 100	
b. Chrysomonadids	100		f. Amoebids	181	
c. Most dinoflagellates	~ 1900		g. Haplosporideans	51	
d. Arcellinids	2000		Subtotal	332	
Subtotal	4100		**Total**	**4832**	**14.7**
Group 3: Meiosis + Mitosis			**III. Secondarily sex lost Protozoa**		
h. A few dinoflagellates	~ 100		a. Euglenids	1000	
i. Heliozoans	100		b. Parasitic flagellates	1900	
j. Opalinates	200		c. Amoebae	19	
Subtotal	400		d. *Nematopsis* spp	330	
II. Agametic sexualized Protozoa			e. *Tetrahymena*	47	
a. Chloromonadids	500		**Total**	**3296**	**10.0**
b. Euciliates	7800		**Grand total**	**16428**	**49.9**
Total	**8300**	**25.2**	**Sexualized gametic Protozoa**	**15142**	**46.0***

In category IV, sexualized gametic protozoans comprise volvocids, chaonoflagellates, filosians, foraminifers, radiolarians, parasitic rhizopods and sporozoans (e.g. majority of gregarines, eucoccids, haemosporinans and myxosporideans) are sexualized and have the gametogenic mechanism. *On the whole, 16,428 species or 49.9% protozoan are not sexualized; of them, ~ 25% sexualized, ciliates and chloromonadids do not have the mechanism for gametogenesis*. The remaining 15,142 species or 46% protozoans are gametic sexualized with ability for gametogenesis (Table 7.1).

7.2 Vegetative Gametes and Reproduction

Within Category I, Group 1 produce vegetative gametes only by mitotic division of haploids or diploids (e.g. *Nematopsis* spp); the latter produce diploid gymnospores (see Fig. 1.16A). It consists of the (i) cryptomonadids (Fig. 7.1A), (ii) chrysomonadids, (iii) most dinoflagellates, (iv) arcellinids (Fig. 7.1B) and (v) *Nematopsis* type of gregarines (Fig. 7.1C).

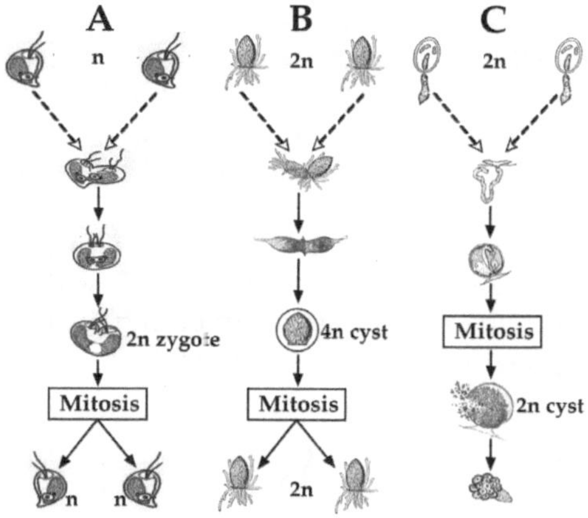

FIGURE 7.1

Mitotic vegetative gametogenesis: (A) haploid cryptomonadid, (B) diploid arcellinid and (C) diploid gymnospores producing *Nematopsis* (based on Figs. 1.7A, 1.9A, 1.16A).

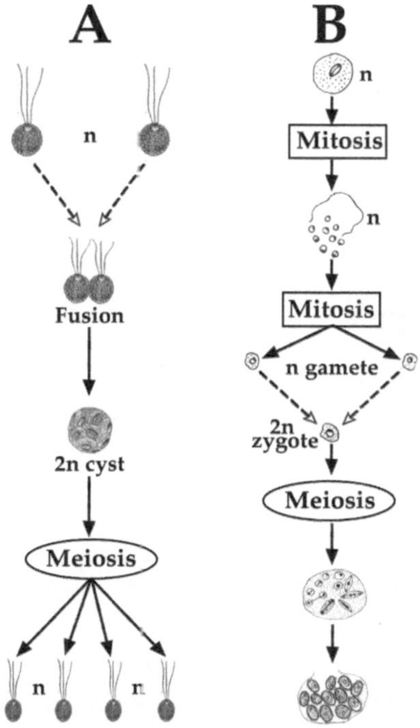

FIGURE 7.2

Mitosis + meiotic vegetative gametogenesis: (A) coccolithophore (see Fig. 1.7B), (B) Haplosporidean protozoans (see Fig. 1.19).

The members of the Group 2 produce vegetative gametes by mitosis followed by 'meiosis'. This group comprises (i) coccolithophores (Fig. 7.2A), *Amblyospora* and (ii) *Haplosporidium* (Fig. 7.2B). Following meiosis, chlamydomonads produce four offspring, microsporideans like *Amblyospora* eight meiospores (see Fig. 1.19) and *Haplosporidium* 32 sporozoites (Fig. 7.2C, see also Fig. 1.20).

Group 3 reproduce through meiosis followed by mitosis. In (v) very few dinoflagellates (Fig. 7.3A), (vi) amoebids (Fig. 7.3B) and (vii) heliozoan (Fig. 7.3C), meiosis is followed by mitosis. In both haploid dinoflagellate and diploid amoebid, eight progenies are produced.

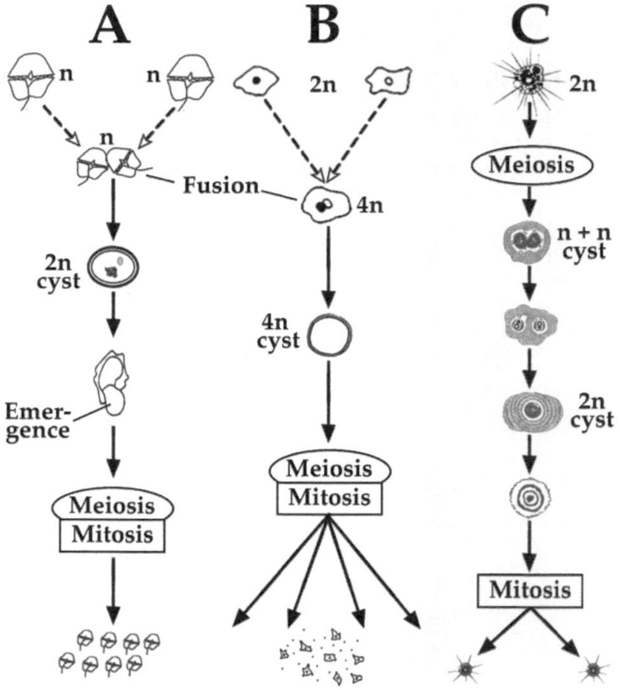

FIGURE 7.3

Meiosis + mitotic vegetative gametogenesis: (A) dioflagellate, (B) amoebid (see Fig. 1.11), (C) heliozoan protozoans (see Fig. 1.9B).

From the survey of 31,570 species (or 96%) protozoan species, for which information is available, 4,832 species or 14.7% protozoans are not sexualized, 8,300 species or 25.2% protozoans are sexualized but do not have the mechanism for gametogenesis and 3,295 species or 10.0% protozoans have lost sex. On the whole, 16,428 species or 50% protozoans are not sexualized and the remaining 15,142 species or 46% protozoans are sexualized gametogenics.

7.3 Germinal Micronucleus and Reproduction

The euciliates are a peculiar taxon, in which the micronuclei act as gametes (see Fig. 1.23). Following meiosis in the diploid micronucleus of a conjugant, four micronuclei are generated, of which three degenerate but one persists, recalling oogenesis in metazoans. The persisting micronucleus divides and yields two pronuclei. One pronucleus remains immotile in each conjugant and the other migrates to the conjugating partner and fuse with the immotile pronucleus of the partner to form a synkaryon or a zygote (Fig. 7.4). The fate of the micronuclear products in the ex-conjugant differs. Two and three consecutive mitotic divisions occur in the *Paramecium aurelia* complex and *P. caudatum* complex, yielding four and eight micronuclei, respectively. Of them, 50% micronuclei are differentiated into

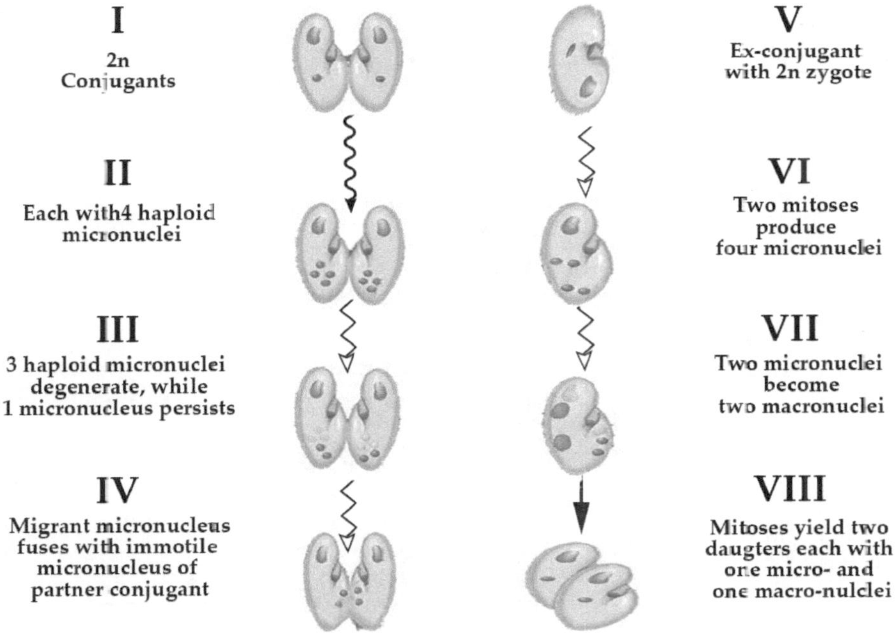

FIGURE 7.4

Micronuclei acting as gametes in conjugating ciliates. → = Mitosis, ∿∿∿▸ = Meiosis (based on Hyman, 1940, Anderson, 1988).

macronuclei. Each ex-conjugant mitotically divides until one micronucleus and one macronucleus are distributed among the daughters, i.e. two progenies appear from two conjugants in *P. aurelia* complex (see also Stephen, 1990) and four from two conjugants in *P. caudatum* complex

(Fig. 7.4). Despite using micronucleus as gamete, the ciliates have achieved almost metazoan-like gametogenesis, albeit complicated by conjugation. Among plants, angiosperms have also achieved metazoan-like gametogenesis, but it is complicated by double fertilization.

7.4 Mitotic Gametes and Reproduction

This category comprises two subgroups. The subgroup 1 consists of the volvocids (Fig. 7.5A). They produce true gametes but mitotically (Fig. 1.8A).

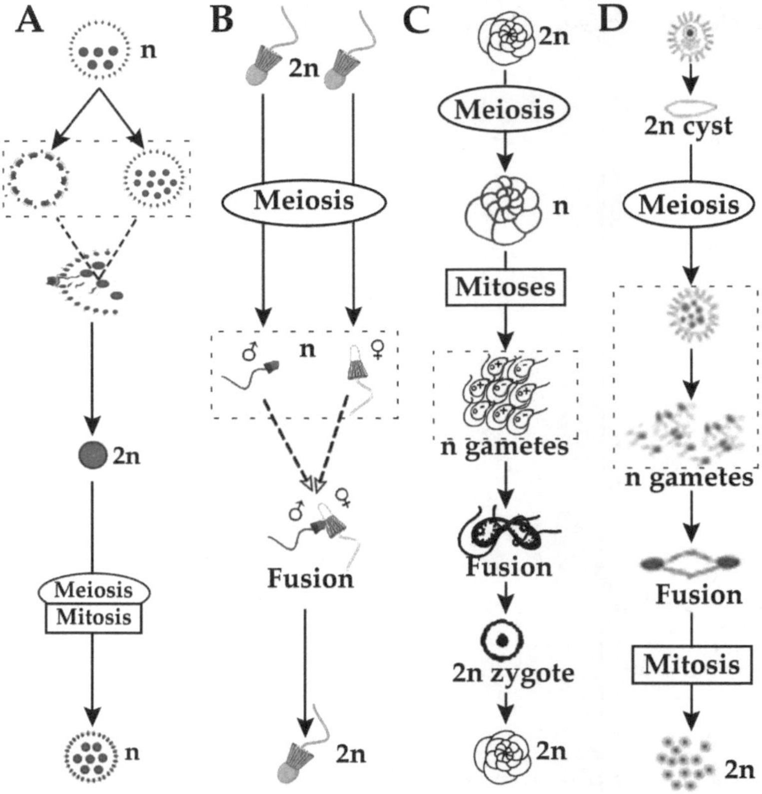

FIGURE 7.5

Life cycles in: (A) *Volvox* (based on Umen, 2020), (B) choanoflagellate (based on Levin and King, 2013), (C) foraminifer (Fig. 1.12) and (D) radiolarian (based on Fig. 1.13). Gametogenic stages are marked by dotted box.

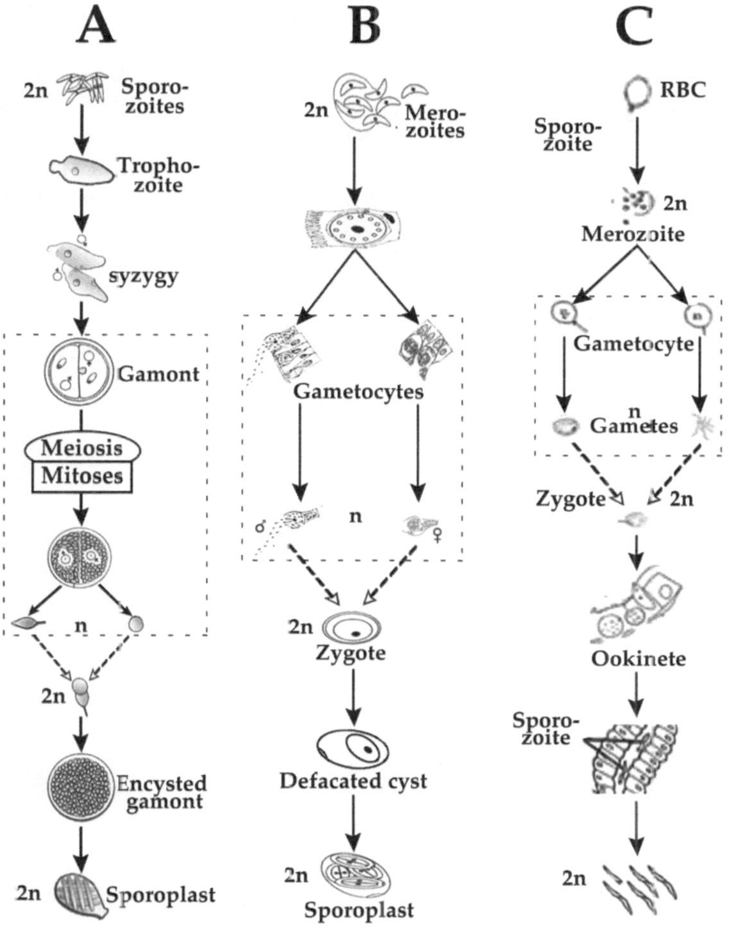

FIGURE 7.6

Life cycles in: (A) *Lecudina*, (B) *Eimeria*, (C) *Plasmodium* (based on Figs. 1.15A, 1.17, 1.18). Gametogenic stages are marked by dotted box.

Unlike the other categories, this subcategory includes only 400 species or 1.2% protozoans alone. Hence, *mitotic generation of gametes may not facilitate species diversity.* In subgroup 2, gametogenic sexual reproduction involves relatively a greater number of stages. It occurs in free-living choanoflagellates among the mastigophores (Fig. 7.5B), foraminifers (Fig. 7.5C) and radiolarians (Fig. 7.5D) among the rhizopods as well as in many parasitic sporozoans (Fig. 7.6A). In the latter, it is further complicated by insertion of schizogony and merogony, as in *Eimeria* (Fig. 7.6B), and involvement of two hosts, as in *Plasmodium* (Fig. 7.6C). As a consequence, isogamic eight flagellated gametes (e.g. *Elphidium crispum,* see also Fig. 1.12) or numerous biflagellated gametes (e.g. *Anlacantha scolymantha,* see also Fig. 1.13) are generated. In contrast, numerous anisogamic gametes are produced in parasitic *Lecudina tuzetae* (see also Fig. 1.15A), *Eimeria* (see also Fig. 1.17) and *Plasmodium* (see also Fig. 1.18). What is missing among the Protozoa is the existence of unifying typical metazoan-like meiosis and its conservation among the protozoans.

8

Induction and Morphogenesis

Introduction

In protozoans, clonal or sexual reproduction is induced by a number of factors; among them, (i) body size or age, (ii) quality and quantity of nutrients or prey, (iii) temperature and (iv) metabolic waste(s) may be considered as important. Sexual reproduction and/or encystation may also be induced by one or the other of these factors. As in metazoans, sex in protozoans is also determined by chromosomal mechanism. However, ~ 6,196 species (Phytomastigophora: 4,845 [Table 1.2] + Haplosporea: 51 [Fig. 1.20] + Microsporidea: 1,300 [Fig. 1.19]) or 19% of protozoans are haplontics, i.e. gametogenic lifespan dominates over a relatively shorter duration of sporogenic one. Not surprisingly, the haploid chromosomal mechanism of sex determination in volvocids has attracted much attention. Morphogenesis is defined as the origin and development of a tissue, organ or organism (see Chamber's Dictionary of Science and Technology). As in metazoans, protozoan species are identified by their morphology. However, their morph changes during the ontogeny, especially in sessiles and parasites. For protozoans, not much information is yet available on sex determination and differentiation. This chapter brings together the relevant information on (i) induction of clonal or sexual reproduction, (ii) sex determination in haplontic volvocids and (iii) morphogenesis in sessile and parasitic protozoans.

8.1 Induction of Clonal Multiplication

On induction by one or the other identified factor, the protozoans begin to clonally multiply by division(s) or fragmentation(s). Being an ongoing process, their multiplication is measured as population growth. From culture studies, a fairly a large volume of information is available on induction and sustenance of clonal multiplication in many protozoans as measured by population growth. Some publications on this aspect have reported very valuable information. Nevertheless, they report the increase in various units, for different durations, considering individual or a combination of factors. Hence, this account has chosen to describe the role played by the identified inducing factors in a kind of unifying models.

Body size is the most important internal factor that induces clonal multiplication. As protozoans grow in size, the surface area per unit volume decreases and thereby slows down the exchange of essential material (e.g. oxygen and carbon-dioxide) both internally and externally. To escape from this sort of senescence, fission is resorted. With regard to body size, more information is available for ciliates as measured by doubling time. In them, the size ranges from 2 µm to 2 mm – a range of one thousand times (see Fenchel and Finlay, 2006). In terms of volume too, the range is from 10^2 µm^3 in *Nivaliella* to 10^8 µm^3 in *Bursaria* among colpodids and from 10^3 µm^3 in *Cyclidium* to 10^9 µm^3 in *Ichthyophthirius* (see Lynn, 2010b). Wang et al. (2005) reported the doubling time as a function of body length (also cross-sectional area) of a few ciliates, as listed below: Fig. 8.1A shows that the doubling time, a measure of population growth and sustenance,

Species	Shape	Length (µm)	Doubling time (h)
Cyclidium sp	Circular	26	6
Uronema sp	Circular	28	6–7
Euplotes plicatum	Elliptical	51	5–7
E. vannus	Elliptical	82	22–30
Keronopsis sp	Irregular	222	53–60

increases with increasing body length, indicating that *surface area per unit volume or body size is the most important internal factor that induces and sustains clonal multiplication.*

Within the genus *Euglena*, the ability differs to utilize nitrogen from nitrogenous sources from species to species. For example, *Euplotes pisciformis* utilize peptones but not amino acids. However, *E. anabaena* utilize, and clonally multiply, when provided amino acids, peptones or ammonium salts, but not on provision of nitrates, while *E. gracilis*, *E. klebsii* and *E. stellata* are induced to clonally multiply, when given any of the four compounds as a source of nitrogen. In *Paramecium aurelia* and *P. caudatum*, the bacterial food *Bacilus subtilis* induce faster multiplication and population growth. But it is not edible for the rhizoids *Hartmannella aquarum* and *Valkampfia magna* (see Jahn, 1934). Being the preferred prey, *Rhodococcus fascians* induce and sustain the maximal clonal multiplication of the flagellate *Cercomonas* sp but the unpreferred *Sphingopyxis witflariensis* support the lowest multiplication (Fig. 8.1B). The optimal temperature that induces and sustains maximal clonal multiplication is around 25°C. Low or super optimal temperature supports only slow multiplication (Fig. 8.1C). The nitrogenous waste products of Protozoa contain urea (Hyman, 1940); however, other waste products are not yet identified. Woodruff (1911, 1913) reported species specific negative effect on sustenance of clonal multiplication in *Paramecium, Stylonychia* and *Pleurotricha*. Figure 8.1D shows the possible effect of waste accumulation on sustenance of clonal multiplication in a protozoan.

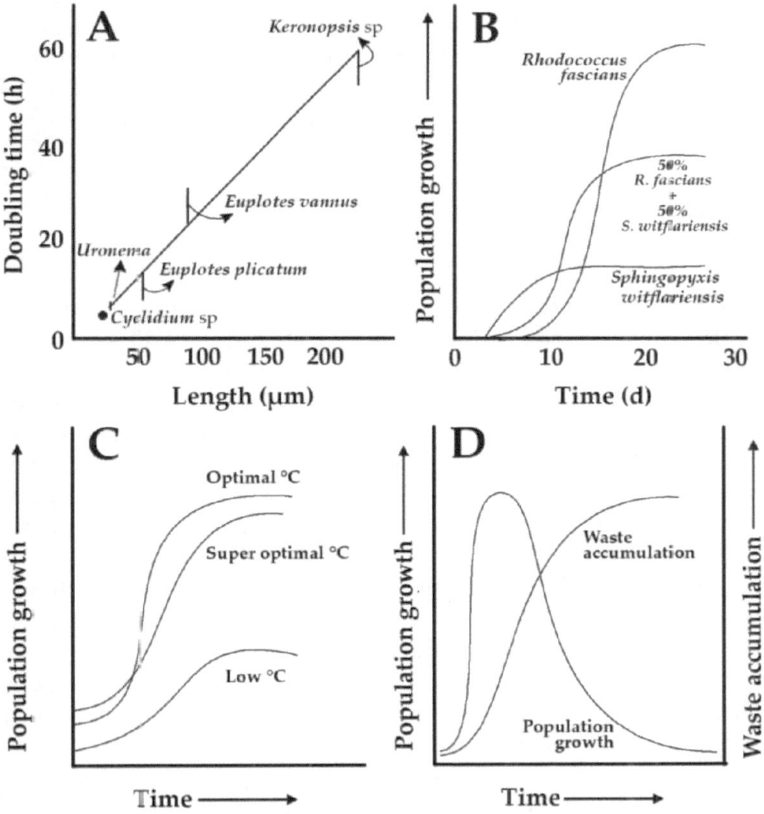

FIGURE 8.1

(A) Doubling time as a function of body length of some ciliates (drawn from data reported by Wang et al., 2005). (B) Population growth of the flagellate *Cercomonas* sp fed on *Rhodococcus fascians* or *Sphingopyxis witflariensis* or combination of them (modified and redrawn from Lekfeldt and Ronn, 2008). Effect of (C) temperature and (D) metabolic waste on population growth of Protozoa (based on Woodruff, 1913, Jahn, 1934).

8.2 Induction of Gametogenesis

In crustaceans, the inactive compound ecdysone and its biologically active metabolite 20-hydroxyecdysone are synthesized from dietary sterols in the Y-organ. Ecdysosteroids play an important role in the control of reproduction. It is in this context, the indication by Hawes (1963) becomes relevant. According to him, 'low doses of ecdysone induce gametogenesis and high doses more of sexual reproduction in the symbiotic flagellates of roaches including in *Trichonympha* and *Barbulanympha*'. However, computer searches with key words 'protozoan hormone' yielded Sonneborn (1942) alone reporting the effect of culture media on sexual development. There are other publications (e.g. Klein, 2004), which reported how the gender-dependent hormone(s) of the host modulate incidence and intensity of infection by protozoan parasites. Briefly, the scope for production of sex hormone(s) by either free-living or parasitic protozoans seems to be remote.

8.3 Gametocyte Ratio

Recent advances in *Plasmodium* seem to point out the involvement of transcription factor in determination of gametocytic sex. Before it, the high risks encountered of the sporozoites of *Plasmodium* spp may have to be explained. On their release from an intracellular ookinete of the infected anopheline mosquito, the sporozoites encounter enormous risk during their passage through the gut lumen prior to reaching the salivary glands. In *Plasmodium vivax*, for example, 3,700 sporozoites/oocyst are produced, of which only 850 sporozoites/oocyst or 23% sporozoites reach the salivary glands. In *P. falciparum*, this value is also reduced from 3,400 to 650 sporozoites or 19% in *Anopheles gambiae*. However, the corresponding values are as high as 90% in *A. gambiae* in Kenya and 57% for it elsewhere. The loss incurred during the passage via the gut lumen and entry into the salivary gland increases the fraction of sporozoites from 10% to 81% (see Simonetti, 1996). Thus, the reductions vary widely for *Plasmodium* species to species, *Anopheles* species to species and country to country. To compensate this widely varying magnitudes of risk, *Plasmodium* spp modulate the sex ratio of the gametocytes and thereby increase the number of oocytes. For example, *P. falciparum* generates three to four female gametocytes against every male gametocyte. Following meiosis, four spermatids from each male gametocyte but only one ovum from each female gametocyte are produced. As a result, the gametic sex ratio is brought to 1 ♀ : 1 ♂ in the mosquito gut (see Talman et al., 2004). Hence, the sex of gametocytes seems to not depend entirely on the so-called sex chromosome(s) but more on the selective expression of the sex genes like the highly conserved apicomplexan-specific transcription factor Ap1 AP2 – G1 in *P. falciparum* (see Meibalan and Marti, 2017).

8.4 Sex Chromosomes and Genes

As in algae and bryophytes, protozoans like volvocids are characterized by homothallic or heterothallic haploid generation. In them, the differentiated Sex Determining Regions (SDR) are located on male and female U and V chromosomes, respectively. Homothallism or monoecy involves oogamous fertilization, as in the ancestral *Volvox africanus*. The homothallic species have retained the nearly intact female-derived SDR and another SDR containing the acquired male determining gene *MID*. Using *de nova* whole-genome sequences, Yamamoto et al. (2021) have identified a heteromorphic SDR of ca 1 Mbp in male and female genotypes of heterothallic *Volvox reticuliferus* and a homologous region (SDLR) in the closely related homothallic *V. africanus*. They have also found a multicopy array of the male-determining gene *MID* in a different genomic location from the SDLR. Thus, the ancestrally female genotype may have acquired *MID* and thereby gained male traits. The heterothallic volvocids ensure fertilization by (i) oogamy, as in *V. reticuliferus*, *V. carteri* or (ii) anisogamy, as in *Eudorina* or (iii) isogamy with + and – mating types (e.g. *Chlamydomonas reinhardtii*, *Gonium pectorale*, *Yamagishiella unicocca* (see Fig. 8.2). The loss of *MID* is one reason for the transition from homothallism to heterothallism.

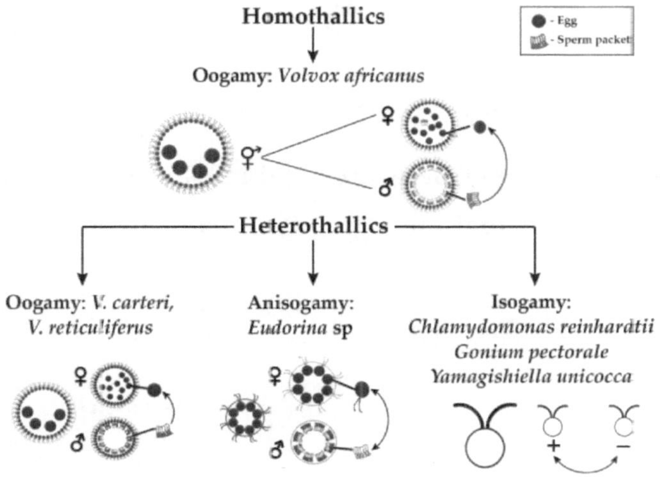

FIGURE 8.2

Patterns of gametes and modes of fertilization in volvocids and chlamydomonadids (modified from Yamamoto et al., 2021).

8.5 Morphogenesis

Though morphology serves as the most important feature for species identification, a species can be monomorphic, as in most hermaphrodites or dimorphic, as in gonochorics. More than 80% metazoan species pass through an indirect life cycle (see Pandian, 2021b) involving one or more larval stages. For example, the free-living penaeid shrimp pass through at least four larval stages namely nauplius, protozoea, mysis and post-larva (see Pandian, 2016). The parasitic digeneans also pass through a minimum four larval stages but some through as many as six stages namely miracidium, sporocyst, redia, cercaria, mesocercaria and metacercaria (see Pandian, 2020). The morphic changes cost resources and time. Most metazoans characterized by indirect life cycle are polymorphics and have the resources to pay the cost of morphic changes during the ontogenesis. In protozoans, unicellularity may not allow the storage of adequate resources to meet the cost of changing their morph both during clonal multiplication and sexual reproduction. Sexualized (75.6% or 24,422 species including 7,800 ciliate and 500 chloromonadid species) can also ill-afford too many morphogenic changes. However, most protozoans pass through a cyst stage to tide over unfavorable situations. Hence, they can be considered as dimorphics.

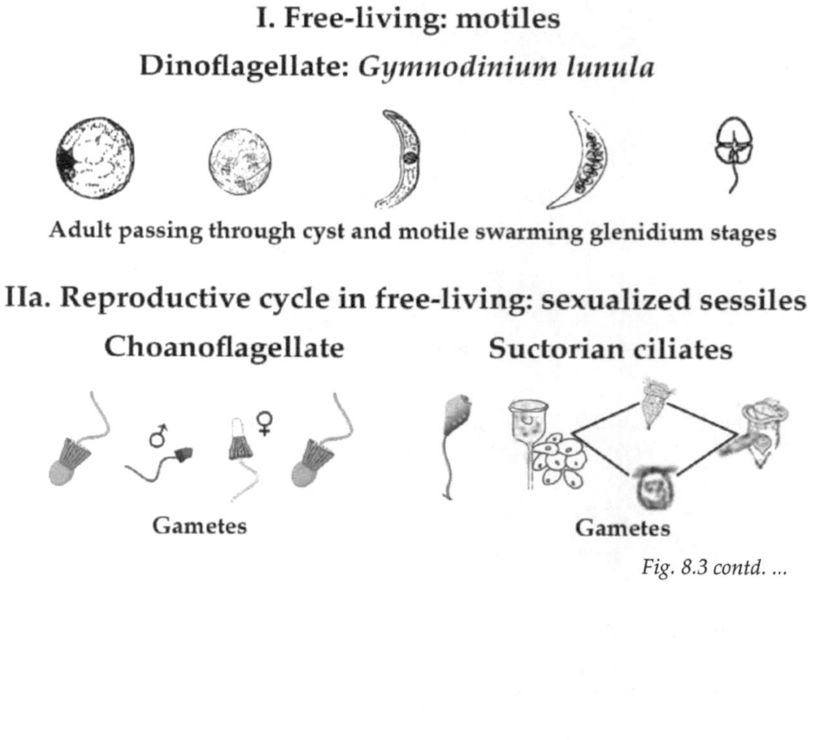

Fig. 8.3 contd. ...

...Fig. 8.3 contd.

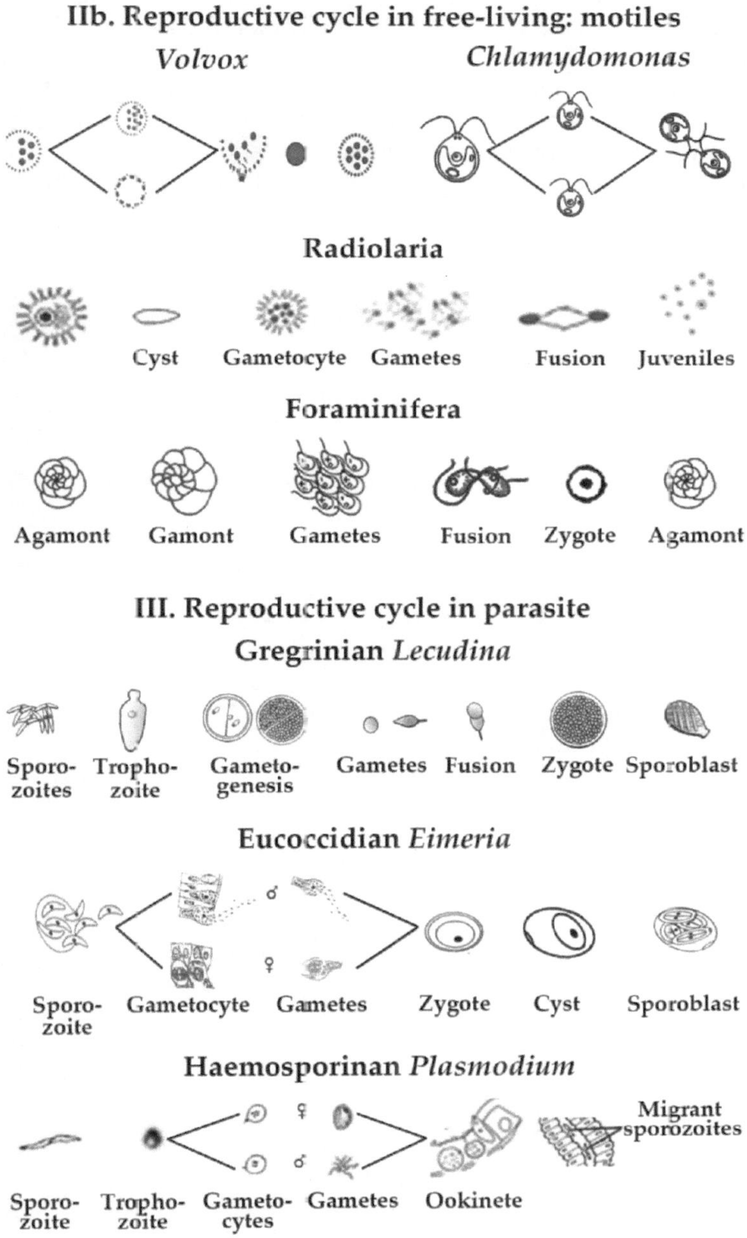

FIGURE 8.3

Morphogenesis during the life cycle of selected protozoans.

The free-living motile dinoflagellate *Gymnodinium lunula* pass through the cyst and motile swarming glenidium stages. However, colonials like *Volvox* and choanoflagellates pass through only the gametic stage (Fig. 8.3). Solitary chlamydomonadids do it through mating types. Radiolarians pass through at least five gametogenic stages. Characterized by heterogonic life cycle, the foraminifers pass through micro- and macro-spheric stages, besides gametogony. With involvement of gametogony, sporogony and schizogony, parasitic sporozoans are more polymorphics than free-living protozoans (Fig. 8.3).

9

Diversity: Symmetry-Sexuality-Motility

Introduction

A major objective of this books is to know whether the unicellularity limits the protozoan species diversity and if so, identify and quantify the factor(s) responsible for it. This account elaborates this aspect in the context of metazoans, for which relevant information has become available (Pandian, 2016, 2017, 2018, 2019, 2020). It has identified the following major factors: 1. Symmetry, 2. Clonality, 3. Sexuality (gonochorism) and 4. Motility.

9.1 Symmetry

Cuvier (1816) was the first to recognize symmetry as an important characteristic for classification of animals. He divided Eumetazoa into two major groups namely (i) Radiata and (ii) Bilatterea. Hyman (1940) considered Porifera as 'asymmetric or radially symmetric Eumetazoa of cellular grade' organization, Cnidaria as 'primary radial or radiobilaterally symmetric Eumetazoa of tissue grade' organization and Acnidaria as 'bilaterally symmetric Radiata'. To these, the secondary radial symmetric Echinodermata of organ/system grade organization may have to be added. Hence, the sub-classification like the Enterocoela and Coelomata may not hold true. Still, two points have to be noted. The echinoderms are unique in that their embryos and larvae display bilateral symmetry but the adults pentamerous radial symmetry. Holothurians exhibit radiobilateral antero- posterior polarity. The existence of radiobilateral symmetry and secondary radial symmetry, i.e. bilateral larval symmetry and adult pentamerous radial symmetry indicates the inability of radial symmetry to assume total coverage of a taxonomic group and entire ontogeny.

The primitive, structurally simpler (< 10 tissue types) Porifera, Cnidaria and Acnidaria are characterized by radial symmetry. Surprisingly, the structurally relatively more complex primitive deuterostome Echinodermata are also radially symmetric, albeit secondarily. Equally surprising is the manifestation of bilateral symmetry in the unicellular Protozoa (see p 3). On the whole, of 1,496,509 major phyletic metazoan species, 26,575 (or 1.8%) and 1,469,934 species (or 98.2%) are characterized by radial and bilateral symmetry, respectively. The former includes 6,644 species/phylum, in comparison to 299,302 species/phylum among the bilaterally symmetric five phyla (Table 9.1). In other words, the number of bilateral species/phylum is ~ 44 times more than that of radiata. Hence, *manifestation of bilateral symmetry has greatly accelerated species diversity. But the radiata has decelerated it.*

Radial symmetry seems to have imposed sessility (in Porifera, Cnidaria and 700 speciose Crinoidea) or sedentary mode of life in the remaining 6,300 echinoderm species. The symmetry ensures 100% clonality or clonal potency, but although the potency is reduced to 1.9% in the structurally complex secondarily radiate echinoderms.

Within radiates, the structurally more organized echinoderms are less speciose than the structurally simpler Porifera and Cnidaria. On four counts, the echinoderms simulate the primitive radiates but deviate from the bilaterals. (i) As in radiates, the gonads of echinoderms ranges from no discrete organ/system in crinoids to filamentous tubules in holothurians (see Pandian, 2018). (ii) Adoption of the intraovarian pattern by the 900 speciose Echinoidea necessiates both oogenesis and vitellogenesis to take place within the ovary. In contrast, the extraovarian pattern is adopted by most bilaterals, in which oogenesis occurs in the ovary and vitellogenesis elsewhere. (iii) Unlike in radiates, echinoderms do not have an intromittent organ. Consequently, most radiates release gametes into water and ensure the costlier external fertilization, within which either male gametes alone are released or both male and female gametes are released. The former saves investment on more valuable eggs. The recalculated values for these two groups from Table 16.4 of Pandian (2021b) listed below show that the mode

Phylum	Release of (%)	
	sperms and eggs	sperms only
Porifera	33.2	66.8
Cnidaria	35.4	64.6
Acnidaria	0.0	100.0
Echinodermata	99.2	0.8
Mollusca	6.2	93.8

of gamete release of echinoderms simulates more of the primitive radiates rather than that of bilaterals like molluscs. (iv) It is known that spatial distribution within a single habitat limits species diversity. Like radiates, the distribution of echinoderms is also limited only to a marine habitat. Interestingly, in the more speciose radiates, Porifera (8,553 species) and Cnidaria (10,831 species), 3% sponges and 0.2% cnidarians have successfully colonized freshwater habitats, whereas the less speciose Acnidaria (166 species) and Echinodermata (7,000 species) are entirely limited to marine habitats alone (see Table 2.1, Pandian, 2021b).

TABLE 9.1

Effect of selected characteristics on species number

Phylum	Species (no.)	Symmetry	Clonality (%)	Gonochorism (%)	Motility
Porifera	8,553	Radial	100.0	38.10	Sessile
Cnidaria	10,856	Radial	100.0	100.00	Sessile
Acnidaria	166	Radial	100.0		
Echinodermata	7,000	Radial	1.9	99.90	2 mm/s
Subtotal	26,575				
Protozoa	32,950	Bilateral	100.0	24.80	~ 4 mm/s
Platyhelminthes	27,700	Bilateral	0.7	0.16	~ 2 mm/s
Annelida	16,911	Bilateral	1.1	76.00	85 mm/s
Mollusca	118,451	Bilateral	0.0	77.10	3 mm/s
Arthropoda	1,242,040	Bilateral	0.0	100.00	11 m/s
Vertebrata	64,832	Bilateral	0.0	99.80	~ 111 m/s
Subtotal	1,502,884				
Total	1,529,459				

Despite unicellularity, the manifestation of bilaterality has enriched diversity to 32,950 species in Protozoa. Irrespective of multicellularity and bilaterality, the acoelomate Platyhelminthes (27,700 species) and the coelomate, segmented Annelida (16,911 species) are less speciose than Protozoa. This requires an explanation. The reason for the limited diversity may be traced to the existence of 99.84% hermaphroditism in Platyhelminthes (see below). Unlike posessing protective shell(s) by the close relatives like the 118,451 speciose Mollusca, almost none of the annelid has a protective shell to escape predation. Being protected by a shell, the Gastropoda have diversified into 90,000 species. In its absence, the diversity is limited to 800 speciose among the highly motile Cephalopoda. Clearly, *the diversity depends more on protection against predation than motility*. Among Protozoa, the most speciose Rhizopoda (11,550 species), known for its slowest motility of

2–3 µm/s, Foraminifera (4,500 species) and Arcenellida (2,000 species) are protected by a shell. Some dinoflagellates are also enveloped and protected by a coat of gelatinous or tectinous material. Not surprisingly, protozoans are more speciose than the annelids.

9.2 Clonality

To gain new gene combinations – raw material for evolution and speciation, organisms depend on (i) random mutation, (ii) segregation during meiosis and (iii) recombination at fertilization, that brings together the whole sets of haploids chromosomes from two individuals. As clonals agametically arise from totipotent stem cells, they may depend on random mutations alone to gain new gene combinations. While the advantageous cloning saves time and resources on progeny production, its chances of gaining new gene combination is limited. Consequently, clonality may reduce the scope for species diversity.

All the 32,950 protozoan species are clonals. In fact, they multiply by cloning, which is rarely interrupted by sexual reproduction. Approximately, 3,300 species or 10.0% protozoans have secondarily lost sex (Table 5.1). For progeny propagation, these sexless protozoans depend only on clonal multiplication. *The ubiquitous and dominant existence of clonality may decelerate species diversity in Protozoa.* More importantly, most protozoans have opted for the costlier mode of fragmentery multiplication, which seems to have imposed severe restrictions on their species diversity. Clonal multiplication may take place either by fragmentation (including binary and multiple fissions) or budding. Fragmenters share the parental body mass. But budders develop new tissues/organs using the parental reserved resources. Hence, fragmentation is costlier than budding. From a survey, Pandian (2021b) found that of 32,519 clonal metazoan species, only 6% are fragmenters, whereas 94% are budders. Among protozoans, budding is limited to the colonial volvocids (352 species) and sessile colonial peritrichids (683 species) and suctorians (233 species). In all, the number of budding protozoans may not exceed 1,000 species (or < 3%). Hence, the remaining 97% (in comparision to 94% metazoan budding species) protozoans are fragmenters.

Incidentally, the 2% echinoderm clonals are all fragmenters. Amazingly, their bilaterally symmetrical larvae are all budders. The adoption of the costlier mode of fragmentation by the 2% adult echinoderms may also have limited their species diversity.

9.3 Sexuality

Within Protozoa and Metazoa, sexuality includes dioecy or gonochorism and monoecy or hermaphroditism. The expression of male and female sexes within an individual results in hermaphroditism but different individuals leads to gonochorism. In gonochores, new gene combinations are generated during meiosis and at fertilization. While gaining the combinations during meiosis, hermaphrodites miss them at fertilization, as gametes arise from the same individual. Consequently, gonochorism may tend to accelerate evolution and speciation but hermaphroditism may decelerate them.

Sex was discovered by organisms ~ 2 BYA and was manifested in animals at different times during the geological past. For example, sex was manifested in foraminiferans 1.0 BYA but 100 MYA in volvocids, as and when they originated (see Fig. 10.3). Surprisingly, 10% protozoans have secondarily lost sex also at different times (see p 140). Hence, only a fraction of protozoans are truly sexualized with the ability for gametogenesis. Among 32,950 protozoan species, only 15,142 species are sexualized and are capable of gametogenesis. Of them, 7,706 and 7,438 species (Table 5.2) or 23.4% and 22.6% are hermaphrodites and gonochores, respectively. *In Protozoa, the restriction of sexualization and gametogenesis to 50%, and limitation of gonochorism (dioecy) to 23% may have decelerated diversity to 32,950 species*. The existence of nearly 100% gonochorism in Arthropoda and Vertebrata has enriched species diversity by two to several times. Notably, with 0.2% gonochorism, the bilaterally symmetrical Platyhelminthes remain less speciose than Protozoa.

9.4 Motility

Motility is the spatial displacement of animals, which creep, crawl, swim, jump, wriggle, walk, run or fly. Unlike symmetry, clonality and sexuality, it is not a unified characteristic but varies widely with regard to many factors. (i) *Body size* significantly influences the speed of motility, which ranges from 2–3 µm/s in Rhizopoda to 400 km/h in Vertebrata and thereby introduces differences in units of measurement. Within a taxonomic group, such as, ciliates, the reported values range from 400 to 2,000 µm/s (see p 58). (ii) *Motility modes*: Many animals have alternate modes of motility. The fish can swimp or leap, tiger can walk or run and thereby render it difficult to reach at a specific value for motility. (iii) *Ontogeny*: Among lepidopterans, caterpillars wriggle but the adults fly. Tadpole larvae swim but the adult frogs jump. (iv) *Motility duration and distance*: Not many polychaete worms and frogs can continue motility for longer than a few minutes. On the other hand, many migratory animals like the penaeid shrimp *Penaeus indicus* continues to swim at a stretch covering a distance of 380 km at the speed of 8 km/d (see Pandian, 2016). The anguillids undertake catadromous migration covering a distance of 4,000 km (see Pandian, 2021b). The migratory Monarch butterfly *Danaus plexippus* can fly over a distance of 3,000 km (Miller et al., 2012). As do migratory birds; some of them fly over a distance of 30,000 km (e.g. Arctic tern, *nationalgeography.org*). (v) *Motility strategy*: Animals engage one or the other organelles/appendages to spatially disperse. Scyphozoans and cephalopods eject water by forcibly pumping water to push them forward – a strategy not used by many animals. But the ctenodrillid polychaetes alter specific gravity to undertake the low cost vertical migration. Use of swim bladder by fish in water and air-sacs by birds in air is well known.

These wide variations question the comparability of values listed in Table 9.1 and consideration of motility as a factor in regulation of species diversity. As mentioned earlier, the possession of a shell by the slowly creeping 90,000 speciose gastropods to escape predation is a more important factor to ensure species diversity rather than the fast moving 800 speciose cephalopods. Nevertheless, motility serves to escape predation and to search for an appropriate mate to ensure sexual reproduction. Hence, it becomes a factor in controlling species diversity. There are some taxonomic groups, in which motility has significantly enriched species diversity. For example, the species diversity remains (~ 250 species, *brittanica.com*) less for the non-flying ratite birds but that of flying birds has increased to 9,788 species. Importantly, species diversity is controlled more by a combination of these factors rather than a single factor. Considering the identified factors, *species diversity seems to be controlled in the following descending order: symmetry > clonality > sexuality > motility.*

10

Past: Emergence of Protozoa

Introduction

Evolution is an ongoing process, and speciation and extinction are its unavoidable by-products. Extinct organisms have left their remnants in the form of fossil fuel by microbes, coal by plants, imprint by soft-bodied plants and animals, amber by plant gum secretion over insects and fossil by skeletonous/shelled animals. They tell us that extinction of organisms is not new or rare, and shall continue to occur so long evolution proceeds. Approximately 6,500 rhizopod species or 20% protozoans belonging to the groups arcellinids (2,000 species) and foraminifers (4,500 species) are fortified by a variety of testate covering or shell. Being soft-bodied animals, the remaining 80% protozoans do not stand a chance of being preserved as imprint or fossil.

10.1 Geological Time Table

Using the carbon dating method, geologists have developed a procedure to estimate the age of earth. Accordingly, the age of earth is considered a little longer than 4 billion years (BYs). The period between those times and pre-Cambrian age is considered under three eons, namely Archean from 4.5 billion years ago (BYA) to ~ 3.5 BYA, Proterozoic from 3.5 BYA to 1.2 BYA and Phanerozoic from 1.2 BYA to 600 million years ago (MYA). Organisms, especially the anaerobic bacteria began to appear some 2.5 BYA. Initially, the oxygenic photosynthetic cyanophyceans and subsequently chlorophytes began to appear and caused the Great Oxygenation event by 2 BYA. Their abundance led to the rapid transformation of earth's atmosphere from essentially anoxic to its present state (Bekker et al., 2004). Around

TABLE 10.1

Geological time table with remarks on origin, evolution and extinction of organisms as well as climate change (from Pandian, 2021b and others)

Epoch	Million Y ago	Remarks
		Era : Palaezoic (600–360 million years ago)
Cambrian	600	**Mild climate** – Appearance and evolution of organismic life – most invertebrate phyla — Prevalence of cryptomonadids, dinoflagellates, polychaete-parasitic gregarines, haplosporeans
Ordovician	500	**Warming climate** – Appearance of vertebrates – Dominance of invertebrates
Silurian	425	***Continents increasingly becoming drier*** –Appearance of land plants and animals
Devonian	405	**Frequent glaciations** – Ascendance of teleosts – Appearance of amphibians and vascular plants – Appearance of Myxosporidea, Trypanosomatina, Haemogregarina
		Era : Mesozoic (359–66 million years ago)
Mississippian	355	**Climate warm and humid** – Many sharks and amphibians – large trees and gymnosperms – Appearance of Opalinata
Pennsylvanian	310	Appearance of reptiles and gymnosperms. Dominance of amphibians and insects – Prevalence of Hypermastigida – Appearance and dominance of Microsporidea
Permian	280	**Widespread glaciations – Atmospheric CO_2 and O_2 reduced – Cooler and drier climate** – Widespread extinction of animals and plants
Triassic	220	**Deserts appear** – Appearance of dinosaurs – Seeded gymnosperms dominant. Extinction of seeded ferns
Jurassic	181	**Rapid evolution of dinosaurs** – Appearance of birds and flowering plants – Appearance of eucoccidian *Eimeria*, Haemosporina
Cretaceous	135	Appearance of monocots, oak and maple forests, and modern grasses and cereals – Massive extinction of dinosaurs
		Era : Cenozoic (since 65 million years ago)
Paleocene	65	Appearance of placental mammals
Eocene	54	**Erosion of mountains** – Appearance of hoofed mammals and carnivores – Prevalence of Polymastigida, euciliatic symbionts – Appearance of Toxoplasmea
Oligocene	36	**Warmer climate** – Appearance of modern mammals and monocotyledons
Miocene	25	Appearance of anthropoids, apes – Rapid evolution of mammals, *Trypanosoma primatum* in apes
Pliocene	11	**Much of volcanic activity** – Appearance of man – Declining forests – spreading grasslands
Pleistocene	1	**Repeated Glaciations – End of Ice age – Warmer climate** – Age of man – Large scale extinction of plant and animal species – Prevalence of Haemosporina, *Giardia, Trichomonas, Entamoeba*

2.0–1.6 BYA, organisms also discovered sex, which was successfully manifested among microbes, plants and animals (Butlin, 2002). Having appeared before 1.6–2.0 BYA, 14.7% non-sexualized protozoans simply engage the transformed vegetative cells as gametes, as in cryptomonadids, coccolithophores, dinoflagellates among the mastigophores and arcellinids among rhizopods (see Table 5.1). It took some more time to discover the atypical meiosis by protozoans, which may occur following mitosis. It took still more time to discover gametogenesis by sexualized protozoans. Notably, the sexualized ciliates have not yet manifested the mechanism for gametogenesis.

The period between 600 MYA and the present is divided into three Paleozoic, Mesozoic and Cenozoic eras (Table 10.1). The Paleozoic era is further divided into four epochs and each of the Mesozoic and Cenozoic era into six epochs. A look at Table 10.1 reveals that 1. The earth began to warm up during Ordovician, continued during the Mississippian and Oligocene (at least once during the Paleozoic, Mesozoic and Cenozoic eras). Hence, climate change and global warming, that are presently encountered, are not new to earth. 2. The extinction of organisms is also not new. In fact, of 80,650 species, 47,700 or 59% protozoans have become extinct (see Table 1.15). 3. Speciation is also not new. Regarding protozoans, the gregarines appeared before the Cambrian epoch and began to evolve and diversify into 1,650 species (Fig. 10.4A). 4. In land, the process of weathering of rocks and landscape formation, which commenced 4 BYA by abiotic factors, was accelerated by bryophytes and lichens from 460 MYA. The landscape with the soil was formed and became available by ~ 300 MYA for occupation by a few terrestrial rhizopods and ciliates.

10.2 Geological Past

Whereas bryophytes, lichens and other primitive plants have greatly contributed to weathering of rocks and landscape formation on land (see Pandian, 2022), the protozoan microfossils serve as reliable sensitive indicators to understand the history of climate changes in the oceans. As mentioned earlier (p 13), the investigations on the isotopic $^{13}\delta$ O : $^{18}\delta$ O ratios between sea water and foraminiferan microfossils have yielded quantitative information on marine climate, especially Sea Level Rise (SLR), Sea Surface Temperature (SST), oxygen content, organic fluxes and so on. The following descriptions are based on Schmiedl (2019).

Microfossils are a dominant constituent of marine deposits. Their diversity, species composition and geochemistry of their shell/skeletal remains from the marine sediment cores narrate the 120 million years history of oceanic climate (Fig. 10.1). The evolution of planktonic Foraminifera parallels that of phytoplankton and reveals their abundance and diversity from the late

Cretaceous to the mid-Miocene. It suggests a close relationship among planktonic ecosystem, SST and SLR. An increase of ~ 300 m for the SLR during the last 120 million years (Haq and Al-Qahtani, 2005) and 4°C for the SST since the last glaciation (see Schmiedl, 2019).

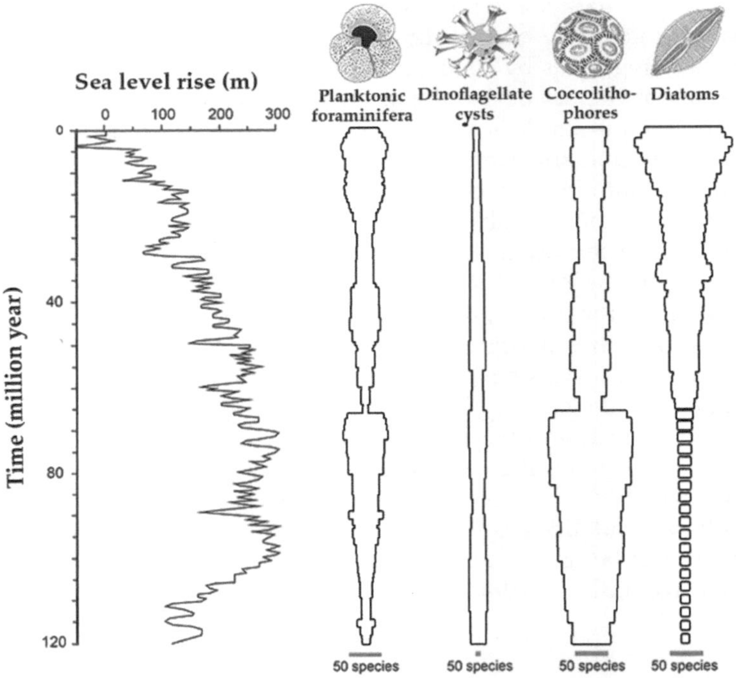

FIGURE 10.1

Evolution and species diversity of a few planktonic taxonomic groups. The diversity of calcareous plankton groups corresponds to major sea-level rise (modified and redrawn from Schmiedl, 2019).

Deep sea benthic ecosystems are linked to the ocean surface through organic fluxes, which serve as the base for food supply to the organisms at the sea floor, and within the sediments. Approximately, 10–40% of the organic carbon produced by phyto- and zoo-plankton is exported from the photic zone but only < 1% of them lands on the sea floor. Microfossils serve as sensitive indicators to monitor the changes in pelagic-benthic coupling through the geological times. Deep-sea ecosystems are influenced by oxygen availability, which is controlled by ventilation of subsurface water masses, water temperature and the microbial oxygen uptake. In this process, benthic foraminifers and their staple $^{13}\delta$ O : $^{18}\delta$ O document the changes in the levels of oxygen and food. Conceptual models suggest that the oxygen level decreases with increasing abundance of food. The critical point that switches from food to oxygen level is located between the mesotrophic and eutrophic conditions (Fig. 10.2A).

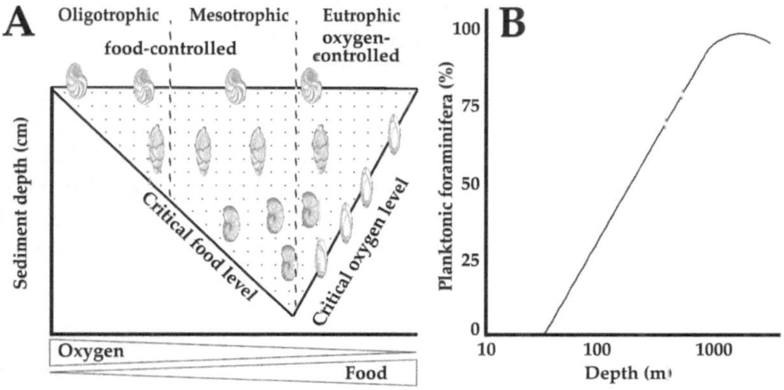

FIGURE 10.2

(A) Conceptual model describing the dependence of benthic foraminifera on food supply and oxygen. (B) Proportion of planktonic foraminifera to total foraminifera in sediments of different seas (modified and redrawn from Schmiedl, 2019).

Quantitative sea level changes can also be estimated from proportions of the planktonic and benthic foraminifers. With increasing depth, the proportion of pelagic foraminifers increases from 50 m depth to 100% at 1,000 m depth (Fig. 10.2B). Reconstruction of the past global SLR have utilized the $^{18}\delta$ O records of deep-sea benthic foraminifers from their pelagic counterparts. Based on regression analysis of planktonic microfossils, the first quantitative reconstruction of global SST distribution for the Last Glaciation Maximum (LGM) was constructed during 1970s and 1980s. Subsequently, the accuracy of SST reconstruction was improved. Based on these efforts, the magnitude of changes suggests an average of 4°C increase from the days of glacial cooling to the present (Schmiedl, 2019).

10.3 Meiosis and Gametogenesis

Although ~ 80% protozoans are soft-bodied and may not have left any imprint or fossil, the incidence of meiosis-gametogenesis as well as symbionts or parasites on fossilizable hosts hint the geological times of the origin and diversity of protozoans. Accordingly, the non-sexualized protozoans that are characterized by mitotically generated vegetative gametes like the cryptomonadids, chrysomonadids, most dinoflagellates (Fig. 10.3A,

Mastigophora) and arcellinids (Rhizopoda, Fig. 10.3B, see also Table 5.1) may have originated at different times during the pre-Cambrian ages. It took some more time for the discovery of atypical meiosis by the other non-sexualized protozoans that are capable of mitotically generate vegetative gametes followed by meiosis, as in coccolithophores, chloromonadids (Fig. 7.2), a few dinoflagellates (Fig. 10.3A), amoebids (Fig. 10.3B) as well as haplosporideans and microsporideans (Fig. 10.4A). The heliozoans (Fig. 10.3B) and protociliates may have taken still some more time to generate vegetative gametes by 'meiosis' followed by mitosis. All the conjugating 7,800 ciliate species that are sexualized but do not have the mechanism for gametogenesis, may have also arisen and diverged between 1.6 BYA and 600 MYA (Fig. 10.4B). The remaining sexualized protozoans may have evolved subsequently; for example, *Volvox* spp have originated just 65 MYA (see Yamamoto et al., 2021).

10.4 Symbiosis and Parasitism

The association between symbionts/parasites and their respective hosts may be considered under three groups. *Group 1 consists of (i) Hypermastigida, (ii) Proteomyxidia, (iiia) Gregarinia, (iiib) Haplosporea and (iiic) Coccidia. Their symbiotic/parasitic association is limited to only one invertebrate host* (Table 10.2). They represent the ancestors of Mastigophora, Rhizopoda and Sporozoa, and

TABLE 10.2

Protozoan symbionts/parasites and their respective invertebrate and invertebrate-vertebrate hosts (compiled from Hyman, 1940, Mehlhorn, 2016, Kinne, 1983a, 1984, 1990 and others)

Parasitic taxon	Examples	Hosts
Group 1: Invertebrate hosts only		
I. Mastigophora		
1. Hypermastigida	*Barbulonympha, Lophomonas, Rhynchonympha*	Cockroaches
	Devescovina, Janickiella, Trichomitus	Termites
II. Rhizopoda		
1. Proteomyxidia	*Plasmodiophora*	Algae, plants
III. Sporozoa		
1. Gregarinia	*Lecudina, Selenidium*	Polychaetes
	Schizocystis, Stylocephalus	Insects
2. Coccidia	*Aggregata* spp	Cephalopod – decapod crabs
3. Haplosporea	*Haplosporidium nelsoni*	Bivalves

Table 10.2 contd. ...

...Table 10.2 contd.

Parasitic taxon	**Examples**	**Hosts**
Group 2: Invertebrate and vertebrate hosts		
I. Mastigophora		
1. Trypanosomatina	*T. gargantua, T. murmanensis*	Fish + leeches
	T. rotatorium	Frogs + leeches
	T. melophagium, T. theileri	
	T. brucei, T. equiperdum, T. evansi	Livestock + insects
	T. gambiense, T. rhodesiense	Human + insects
	Leishmania donovani	Human + insects
II. Rhizopoda		
1. Piroplasmea	*Theileria equi*	Equines, ticks
III. Sporozoa		
1. Haemogregarina	*Haemogregarina bigemina*	Fish + leeches, crustaceans
2. Microsporidea	*Amblyospora*	Arthropods
3. Haemosporina	*Plasmodium* spp	Human + *Anopheles*, birds + flea
Group 3: Vertebrate hosts only		
I. Mastigophora		
1. Polymastigida	*Giardia* spp	Vertebrates
	Trichomonas faetus	Cattle
	T. hominis, T. vaginalis	Human
	Giardia vaginalis	Human
II. Rhizopoda		
1. Entamoebida	*Entamoeba blattae**	Cockroach
	E. ranarum	Frogs
	E. coli, E. histolytica, Iodamoeba butschlii	Human
IV. Sporozoa		
1. Eucoccidia	*Eimeria*	Birds
2. Myxisporidea	*Myxobolus, Myxidium*	Fish
3. Toxoplasmea	*Toxoplasma gondii*	Felids + birds/rats
V. Ciliophora		
1. Protociliata	*Opalina ranarum*	Frogs
2. Euciliata	Butschliidae, Isotrichidae, *Trichostomatia*	Ruminants
	Blespharocorys	Horses
	6 genes in 3 families	Elephants

* May be an exception

may have been prevalent on the identified hosts since the pre-Cambrian ages. *Group 2 comprises (i) Trypanosomatina, (ii) Piroplasmea and (iiia) the eucoccidian Haemogregarina, (iiib) Haemosporina and (iiic) Microsporidea representing Mastigophora, Rhizopoda and Sporozoa, respectively*. In them, the members engage two hosts, one from an invertebrate (e.g. leech or arthropod), and the other, a vertebrate. They may have originated during the Devonian epoch, when the teleosts arose and became abundant (see Table 10.1). The trypanosomes are an interesting group; their incidence commences from leech–teleost/frog through reptiles and birds to sanguivorous insects–mammals including apes and humans (Table 10.2). Notably, rhizopods with the lowest representative parasitic protozoans (only 250 species, see Table 4.6) includes the least speciose Piroplasmea.

Group 3 includes (i) the mastigophoran Polymastigida, (ii) the rhizopodan Entamoeba spp, *(iiia) Eucoccidia, (iiib) the sporozoan Myxosporidea, (iiic) Toxoplasmea as well as (iva) protociliates parasitic on frogs and (ivb) Euciliates, symbiotic in ruminants, equines and elephants* (Table 10.2). Most members of this group must have emerged only after the Palaeocene (65 MYA), when placental mammals appeared (see Table 10.1). Exceptionally, the protociliates, parasitic on frogs, may have originated during the Mississippian (355 MYA), when the amphibians arose. However, the symbiotic euciliates may have become diverged only after the Eocene (54 MYA), when the hoofed mammals emerged. The entameobids are an interesting group, as their incidences are known from frogs (e.g. *Entamoeba ranarum*) to humans (*E. coli, E. histolytica*).

10.5 Origin and Diversification

A combination of information derived from meiosis–gametogenesis and association between symbionts/parasites–respective hosts provide a greater insight into the origin and diversification of Protozoa. The required information to describe the relationship between the species number and geological time was assembled for Mastigophora (*onezoom.org*, Zakrys et al., 2007, King et al., 2008, Yamamoto et al., 2021) and Rhizopoda (*onezoom.org, ucl.ac.uk, bgs.ac.uk, nhm.ac.uk*). For Sporozoa, the required data were drawn from Table 1.4 and Table 10.1. For the monophyletic Ciliophora, relatively less information is available. Jiang et al. (2018) reported valuable information for species number and time of origin for Ciliophora, peritrichid ciliates, and solitary (e.g. *Stentor*, 1,500 MYA, *Thuricola*, 320 MYA, *Epistylis*, 200 MYA) and colonial (e.g. *Opercularia*, 300 MYA, *Zoothamnium*, 150 MYA,

Vorticella, 130 MYA) peritrichids. The assembled information for the four major taxonomic classes was grouped under four groups. 1. Non-sexualized vegetative gametogenic protozoans, 2. Sexualized but non-gametic ciliates and chlamydomonads, 3. Secondarily sex lost protozoans and 4. Sexualized and gametogenic protozoans. The major Group 1 is further divided into three subgroups: Subgroup (i): The member of this subgroup generates vegetative gametes through mitosis alone, Subgroup 2 consists of taxa, which also generate the gametes through mitosis but is followed by meiosis and Subgroup 3, in which the vegetative gametes are generated through meiosis, followed by mitosis. Analysis of the assembled information, on illustrations in Fig. 10.3 and 10.4, led to draw the following new findings: The subgroup 1 within Group 1 consists of (i) cryptomonadids (100 species), (ii) chrysomonadids (< 100 species), (iii) coccolithophores (< 100 species), (iv) dinoflagellates (~ 2,000 in phytomastigophores, (v) arcellinids (2,000 species) in rhizopods, and (vi) haplosporeans (51 species), that parasitize the marine bivalves and (vii) microsporideans in sporozoans. In the 2,000 speciose dinoflagellates, the distinct nuclear fusion between the two participants may have generated new gene combinations. But evidence for a similar nuclear fusion is wanting for the less speciose cryptomonadids. The reason for the rich species diversity in arcellinids is not yet known. *All these protozoan taxonomic groups may have emerged at different times during the pre-Cambrian age, prior to the discovery of sex by 1.6–2.0 BYA* (Fig. 10.3A, B, Fig. 10.4A). Subgroup 2 consists of the less speciose (i) coccolithophores (< 100 species), (ii) a few dinoflagellates (~ 100 species), in which the vegetative gametes are also mitotically generated but is followed by meiosis. They have also appeared during the pre-Cambrian times (Fig. 10.3A). The same holds true for subgroup 3, in which the vegetative gametes are generated by 'atypical meiosis', which may be followed by mitosis, as in 100 speciose heliozoan (Fig. 10.3B) but not in the 200 speciose protociliates (Fig. 10.4B). *Notably, the atypical meiosis may have been discovered and manifested prior to the discovery of sex. These two processes are, though interrelated, distinctly different and independent.*

Group 2 consists of the 500 speciose chloromonadids and 7,800 speciose euciliates. In them, either the differentiated vegetative motile zoospores, in the former, or the micronuclei in the latter serve as gametes. Descriptive accounts are provided on their sexual reproduction in p 99, 113 and Figs. 7.1, 7.4. Notably, most euciliates are monoecious, whereas the chloromonadids can be monoecious or dioecious; the mating types can be distinguished functionally, if not morphologically. In all, they consist of 8,300 species or 25% of protozoans. They must have originated after the discovery of sex and perhaps between 1.6 BYA and the pre-Cambrian epoch (Fig. 10.4B).

Group 3 consists of protozoans that have secondarily lost sex with or without gametogenic mechanism at different geological times (see below).

1. Euglenida	600 MYA	5. *Amoeba* spp	500 MYA
2. Symbionts in roaches + termites	500 MYA	6. *Nematopsis* spp	400 MYA
3. *Trypanosoma* spp	400 MYA	7. *Tetrahymena* spp	400 MYA
4. *Giardia* spp, *Trichomonas* spp	10 MYA	8. Symbionts in hoofed mammals	50 MYA

In all, 3,296 species or 10.0% protozoans (see Table 5.1) have secondarily lost sex. Four different taxonomic groups within Mastigophora have lost sex during the geological past from 600 MYA to < 10 MYA. The reason for the loss of sex is traceable to polyploidy in *Amoeba* spp (see Hawes, 1963). It is not known whether 1,000 speciose Euglenida have diversified prior to the loss of sex. In fact, this group, in which sex is secondarily lost, is brought to light and highlighted for the first time in this account. More research is required to know the causes for the loss of sex in them.

In Group 4, the members are characterized by true sexual reproduction along with mechanism for gametogenesis. In them, the life cycle is diverse and complicated.

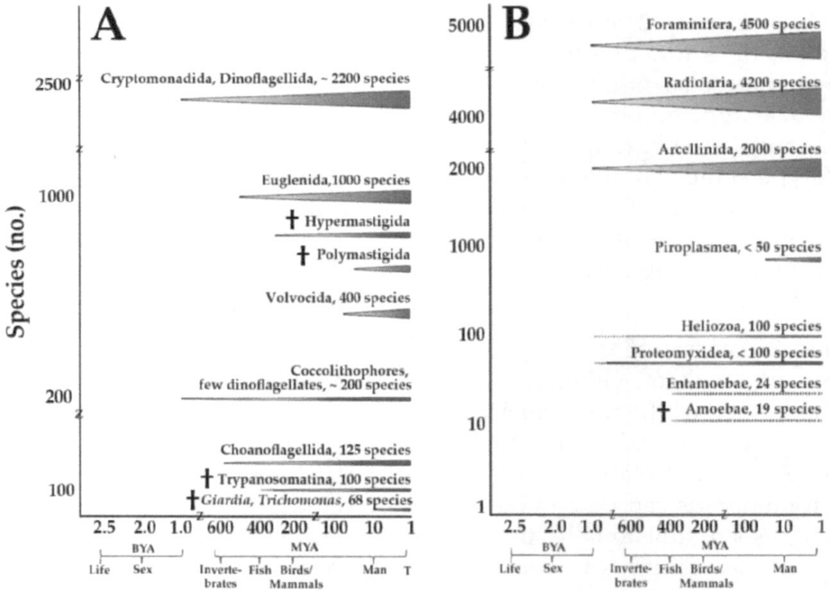

FIGURE 10.3

Possible course of origin, evolution and species diversity in some taxonomic groups of (A) Mastigophora and (B) Rhizopoda. ↑ = diversification to insects ~ 300 MYA, † = secondarily lost sex.

They may be divided into two subgroups: The free-living subgroup consists of (i) volvocids, (ii) choanoflagellates, (iii) foraminifers and (iv) radiolarians. The second subgroup includes the parasitic sporozoans. In the colonial flagellates, gametogenesis takes place through mitosis in volvocids but by meiosis in choanoflagellates. Their diploid zygote undergoes meiosis and repeated mitoses to form the colony in volvocids (see Fig. 7.5A) but it is limited to mitoses alone in choanoflagellates. Colony formation is an ancient discovery, originated 200 MYA and 600 MYA in volvocids and choanoflagellates, respectively (Fig. 10.3A). This is also true for ciliates. In peritrichid ciliates, solitaries have reported to originated since 1,500 MYA but the colonials 300 MYA (see Jiang et al., 2018). Among solitaries, radiolarians

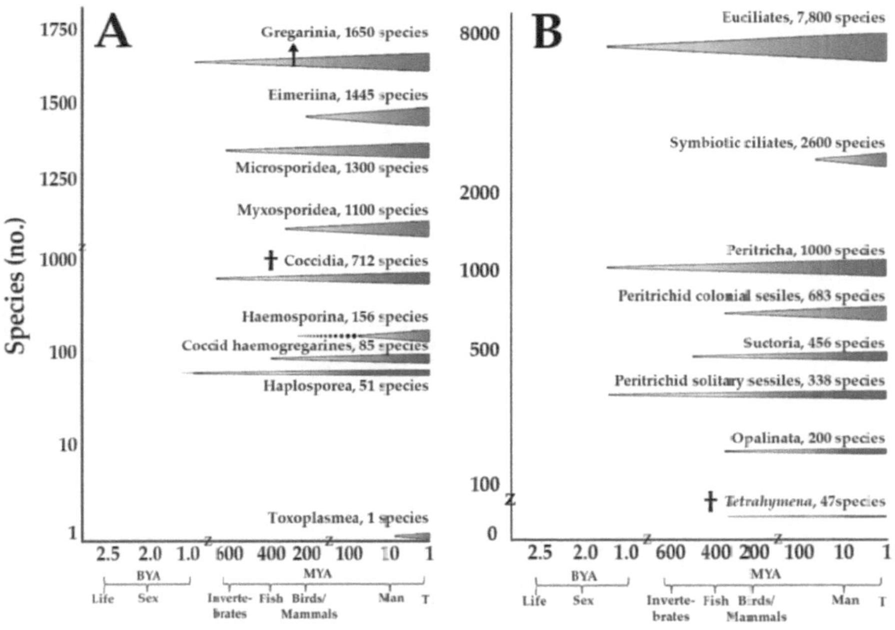

FIGURE 10.4

Possible course of origin, evolution and species diversity in different taxonomic groups of (A) Sporozoa and (B) Ciliophora. † = secondarily lost sex.

undergo meiosis followed by repeated mitoses. With sexual reproduction generating new gene combinations (Fig. 10.3B), they have originated around 1.0 BYA and diversified into 4,200 species. The heterogonic life cycle of the 4,500 speciose foraminifers involves the obligate alternation of gamogony with clonal schizogony. The foraminifers are the most diversified taxonomic group among protozoans, which have undergone the fastest diversification into 4,500 living species, since their origin by 1,000 MYA (Fig. 10.3B). With intriguing sexualization without gametogenic mechanism, the ciliates originated around 2 BYA, when sex was discovered. As indicated earlier, the manifestations of sex and gametogenesis are independent processes. Peritrichids have a relatively short history. In them, solitaries originated 1.5 BYA and diversified into 338 species but the colonials, which appeared ~ 320 MYA diversified faster into 683 species (see also Table 1.12).

11

Present: Conservation and Dormancy

Introduction

The negative anthropogenic activities have driven many organisms to a point of near extinction. Presently, efforts are being made to conserve them. *In situ* conservation demands legislation and its implementation by governments but it may be of local and temporal importance. The engraved stone edifices of the Indian Emperor Asoka (276–232 BC) revealed that he was the first to ban inland fishing operation duration spawning season. Employing scientific indices, IUCN, an independent organization under the umbrella of UNESCO, has recognized some species as red listed endangered plants and animals, and demanded urgent and strict implementation of laws requiring different measures of conservation. For obvious reasons, the IUCN hitherto has not recognized any protozoan species as endangered, albeit 47,700 protozoan species have become extinct in the geological past (Table 1.15). On the other hand, *ex situ* conservation of endangered organisms requires scientific techniques, which may be of global importance. There are reports for maintaining *Entosiphon sulcatum* (see Hyman, 1940) and *Paramecium aurelia* (Woodruff, 1926) for over 947 and 1,500 generations, respectively. Though techniques are described for maintaining parasitic protozoans (e.g. Ahmed, 2014), neither a protozoologist nor an institution seems to maintain a culture of Protozoa. In fact, no museum around the world is known to maintain representative specimens for Protozoa. Besides from these, protozoans have developed their own strategies like encystment and/or dormancy to conserve themselves for a short or longer duration. In this chapter, surveys are made to identify protozoans that do not encyst and those that produce cysts, in which protozoans can survive for longer durations.

11.1 Dormancy – Excystation

To tide over unfavorable conditions in ephemeral freshwater systems like ponds, puddles, ditches and streams, the need for cyst formation is understandable. But it is difficult to comprehend the need for it in marine habitats. Yet, many animals do it in coastal waters (Pandian, 2016, 2021a). It was initially suggested that with decreasing cyst size, the survival duration of life is prolonged (Pandian, 2021b). However, it is the prevention of light and consequent darkness that induces and sustains dormancy over a longer duration (Pandian, 2022). Following subitaneous encystation, most protozoans excyst or emerge from the cyst within a short period. In others, which are buried deep in sediment or stored in darkness, dormancy is induced. Dormancy is a life history strategy involving suspension of active life, and reduced or suspended metabolic activity, mediated by exogenous factors (Ross and Hallock, 2016). It is defined as the inability of a cyst not to excyst or emerge under unfavorable condition. It allows the persistence of cysts of a genotypes, population or species to (i) survive after the death of parent organisms (e.g. foraminifers *Ammonia beccarii, Haynesina germania,* see Ross and Hallock, 2016) and (ii) distribute the genetic diversity through space and time (Long et al., 2015). Dormancy and excystation are closely linked traits with a great impact on survival and sustenance of a species.

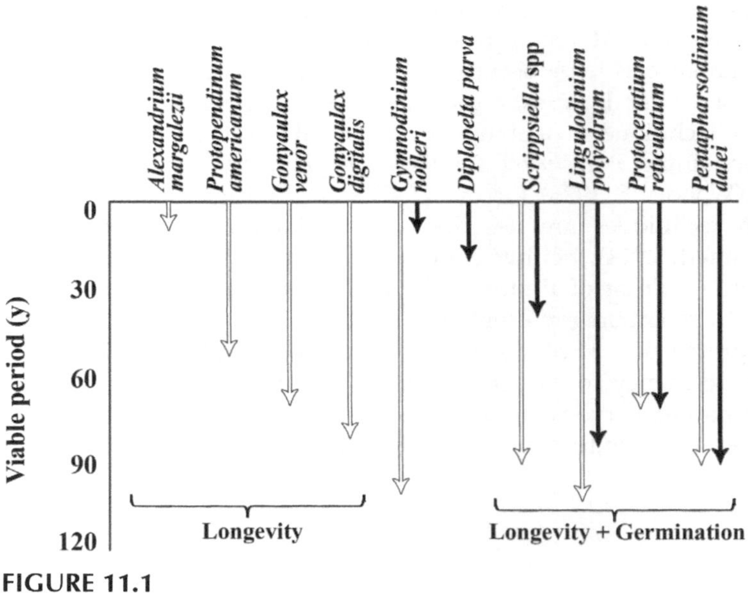

FIGURE 11.1

Survival and germination of dinoflagellate cysts (simplified and redrawn from Lundholm et al. 2011).

Dormancy may ensure survival of life within a cyst over a long period of time (e.g. the longevity) but may not assure the emergence throughout the dormant duration (e.g. longevity cum emergence), for want of resources to facilitate the emergence being followed by multiplication. For example, the emergence success for the stored dinoflagellates is 3% for 9-years old cysts, in comparison to 97% for 1-year old ones (Lewis et al., 1999).

Marine protozoans: Dating of dinoflagellate cysts collected from the marine core sediments is measured by ^{210}Pb activity in gamma spectrometry. Limited information is available on the longevity and longevity cum emergence from cysts of Protozoa. For dinoflagellates, Lundholm et al. (2011) reported valuable data for these durations for the dinoflagellate cysts that were collected from the Swedish deep Fjord. The longevity of dormancy ranges from 10 years in *Alexandrium margalezii* to > 100 years in *Gymnodinium nollerii* (Fig. 11.1). The values are > 90 years for the longevity of dormancy but < 40 years for the longevity cum emergence in *Scrippsiella* spp. However, they are equal for both longevity and emergence with 75 years in *Protoceratium reticulatum* and 90 years in *Pentapharsodinium dalei* (Fig. 11.1). Describing the prevalence of dormancy in organic thick-walled agglutinated spherical cysts of foraminifers, Ross and Hallock (2016) narrated more of cyst resistance to experimental exposure to temperatures and other factors. For example, the cysts of *Reticulomyxa filosa* can successfully withstand complete dehydration, freezing and pressure for 32 days and yet revive within 2.3 days. A more relevant but incomplete information is available for the emergence from < 32 µm sized cysts of *Ammonia beccarii* and *Haynesina germania* after a period of 4 months storage from sealed covers kept in dark, cold containers. Unfortunately, the age of these cysts collected from the core sediment deeper than 140 m depth was not dated. The longevity cum emergence duration for the foraminifer cyst can also be 100 years, as in dinoflagellates. This 100-years for the protozoan cyst are comparable to 125-years for the ephippia of daphnids but not with 320-years for copepods (see Pandian, 2021b). Hence, it is likely that *the cysts of marine protozoans have longevity cum emergence for the period of ~ 100 years.*

Freshwater/parasitic protozoans: Limited information is available for viability (longevity cum emergence) of cysts of free-living and parasitic protozoans. The viability duration ranges from 9 months (mo) in *Euglena gracilis* to 5 years in *Colpoda cucullus* (Table 11.1). For parasitic protozoans, it ranges from 1 year in *Entamoeba histolytica* to 24 years in *Acanthamoeba* sp. Notably, the viability is determined by temperature and moisture content. It is reduced from 24 years at 4°C in *Acanthamoeba* to 24 months *A. culbertstoni* at room temperature, i.e. 20°C. In *C. cucullus*, it is 5 years, when the cysts are exposed alternatively to

TABLE 11.1

Reported viability of freshwater and parasitic cysts

Taxon	Viability	Reference
Free-living protozoans		
a) *Euglena gracilis*	9 mo at 20°C	Strauch et al. (2017)
b) *Tillina magna*	4 mo kept dry	Corliss and Esser (1974)
c) *Colpoda cucullus*	5 y	Corliss and Esser (1974)
Parasitic protozoans		
a) *Entamoeba histolytica*	1 y at 27°C	Tulane University Digital Library
b) *Acanthamoeba costellanii*	8 mo at 4°C	Gupta and Das (1999)
c) *A. culbertstoni*	24 mo at 20°C	Gupta and Das (1999)
d) *Naegleria fowleri*	8 mo at 10–15°C	Biddick et al. (1984)
e) *Acanthamoeba*	24 y at 4°C	Mazur et al. (1995)

natural moisture and drying conditions. But it is reduced to 4 months in *Tellina magna* cysts kept under dry conditions. Incidentally, Hyman (1940) indicated the viability of *Colpoda* cyst from 38 years-old soil and 49 years-old cysts of *Monas, Bodo* and *Cercomonas. E. gracilis* secrete a mucilaginous protective coat and replaces intracellular water content with disaccharide trehalose. It may be interesting to know the property of trehalose in maintaining the stability of the cyst. In *Artemia* cyst, the glycogen conversion to trehalose and reduces glucose to form a variety of insoluble 'melanoidens'. Besides, trehalose is not susceptible to aminolyses. These properties of trehalose ensure greater stability and integrity of the cyst (see Pandian, 1994).

12

Future: Climate Change

Introduction

Due to anthropogenic activity, the earth and its organisms are being more frequently threatened by bacterial epidemics and viral (e.g. COVID-19) pandemics, the serotonin-induced swarming of desert locusts and environmental factors like violent earthquakes, tsunamis, cyclones, storms and pollutants. But these are all transient localized episodes. There is no historical evidence to show that any of these factors either singly or in combination have wiped out an animal species, *albeit* they may drastically reduce population size of the affected taxa. Contrastingly, climate change is a long-lasting phenomenon that covers the entire earth. Not surprisingly, researches on the unprecedented increase in atmospheric carbon dioxide (CO_2) concentration and consequent global warming and ocean acidification during recent years have become the hottest research area. This chapter intends to present an emerging scenario in the context of species diversity.

12.1 Air – Water Interaction

The earth is surrounded by atmosphere consisting of ~ 78% nitrogen, ~ 21% oxygen, 0.03% carbon dioxide (CO_2) and others. With the advent of industrial era in 1750 and the accompanied ever increasing energy extraction from fossil fuels has increased greenhouse gas (GHG) emission and concentrated the atmospheric GHG. In turn, the concentration has led to global warming and ocean acidification, which are collectively known as climate change. Global warming has also begun to melt the polar caps, which leads to rise in sea levels and submergence of coastal areas in seawater. Since 1750, the level of atmospheric CO_2 has risen from 280 ppm to 410 ppm in 2019 and is predicted to go up to 550 ppm by 2050 (Table 12.1). During the last 250 years, the levels of other greenhouse gases have also increased from 715 ppb to 1,866 ppb for methane and from 270 ppb to 332 ppb for nitrous oxide (IPCC, 2021). As a consequence, the global mean temperature increased at the rate 0.2°C/decade over the last 30 years. Most of the added energy is absorbed by waters of the oceans (up to 700 m depth), where temperature increased by ~ 0.6°C over the last 100 years and is continuing to increase (see Pandian, 2015). The available database (Emergency Events Database: *emdat.be*) indicates that the frequency of events like drought, storm and flood increased from ~ 100/y during 1990 to > 200 in 2016; of them, the frequency of drought alone increased from ~ 50 time/y in 1990s to > 120 time/y in 2016.

TABLE 12.1

Changes in climate features during the last 10–30 years and predicted changes by 2050 (from Pandian, 2015), * by 2080s

Climate features	Last 10–30 years	By 2050
Atmospheric CO_2 (ppm)	385	550
pH of oceanic waters (unit)	–0.1	–0.1 to –0.3
Sea surface temperature (°C)	+0.4	+1.5
Coral bleaching (time/y)	+2	+15–25
Sea level rise (mm/y)	1	8*
Hypoxic aquatic system (no.)	400	680
Wind speed %/1°C increase	3.5 %/1°C	Increases

Besides absorbing atmospheric temperature, oceans also absorb CO_2, as it combines with water chemically. Covering 70% of the earth's surface and holding 97% of its water, they serve as a buffer to CO_2 concentration. Consequently, the daily uptake of atmospheric CO_2 by the oceans is 22 million metric ton (mmt). Since the advent of the industrial era, the oceans have absorbed 127 billion metric ton (bmt) carbon as CO_2 from atmosphere. The CO_2 absorbed by oceans ranges between 25 and 40%, i.e. a third of atmospheric carbon emission. Without this 'ocean sink', the atmospheric CO_2 concentration would have by now increased to 450 ppm and a consequent increase in temperature on land.

Hydrolysis of CO_2 increases the hydrogen ion (H^+) concentration with concomitant reduction in pH and carbonate (CO_3^{-2}) concentration. This process of reducing sea water pH and concentration of carbonate ion is called 'ocean acidification'. Consequent to the acidification process, the mean pH level of the world's oceans has declined by 0.1 unit and 0.3 unit reduction is expected by 2050. The decrease in sea water pH and carbonate ion concentration is one of the most persuasive environmental changes in the oceans and poses one of the most threatening challenges to marine organisms. The progressive reduction in the availability of carbonate ion (CO_3^{-2}) renders the acquisition of biogenic calcium carbonate ($CaCO_3$) by calcifying organisms energetically costlier, but may not totally inhibit the acquisition. In fact, the reduction in pH is more critical for the calcifying poikilothermic organisms than the increase in sea water temperature (see Pandian, 2015).

With climate change and consequent melting of polar caps, rise in sea level threatens to engulf most of our coastal cities along the coastal zone. It is in this context, studies on the estimation of sea level rise have become important. Interestingly, some protozoans serve as sensitive indicators of the sea level rise. The foraminifers and testate amoebae are key constituent of microfossils and are useful in reconstruction of data on past sea levels, especially in the Holocene since 1880s. These microfossils are valuable sea level indicators, as their modern counterparts are distributed in narrowly defined niches in the intertidal zone at specific level. The sea level raise can be estimated, when an indicator fossil species is dated and its occurrence level (height) is measured, and compared with those of its living counterpart.

The frequency of tidal flooding, subaerial exposure and other associated factors, to which the indicator foraminifer species are subjected determine and limit their vertical distribution range. For example, the tolerance to the flooding and exposure decreases from the hardiest agglutinated foraminifer *Jadammina macrescens* to the lowest tolerant *Miliamina fusca* in the 80 m wide intertidal zone in eastern Maine, USA (Fig. 12.1A). This pattern of foraminifer zonation occurs in tidal marshes around the world from high latitude salt marshes to low latitude mangrove environments (Gehrels, 2002). Studies on linking geological reconstructions and instrumental observations in these environments have revealed the significant ongoing sea-level rise. In a specific innovative investigation, Charman et al. (2010) showed that the sea level has risen up to 0.1 m height during the period from 1880 to 2000 AD in the representative coast of Nova Scotia, Canada (Fig. 12.1B). Interestingly, both the indicators foraminifer and testate amoebae have brought a more or less the same trend, conforming that they are equally sensitive indicators.

12.2 Green Shoots and New Hopes

For protozoans, limited information is available on the effects of climate change. Jahn (1934) indicated that many protozoans can tolerate and grow at pH ranging from 6.5 to 8.0. *Paramecium aurelia* undertake binary fission at almost comparable rate in pH range between 5.9 and 7.7. With progressive acidification, the decrease in sea water pH and carbonate ion concentration

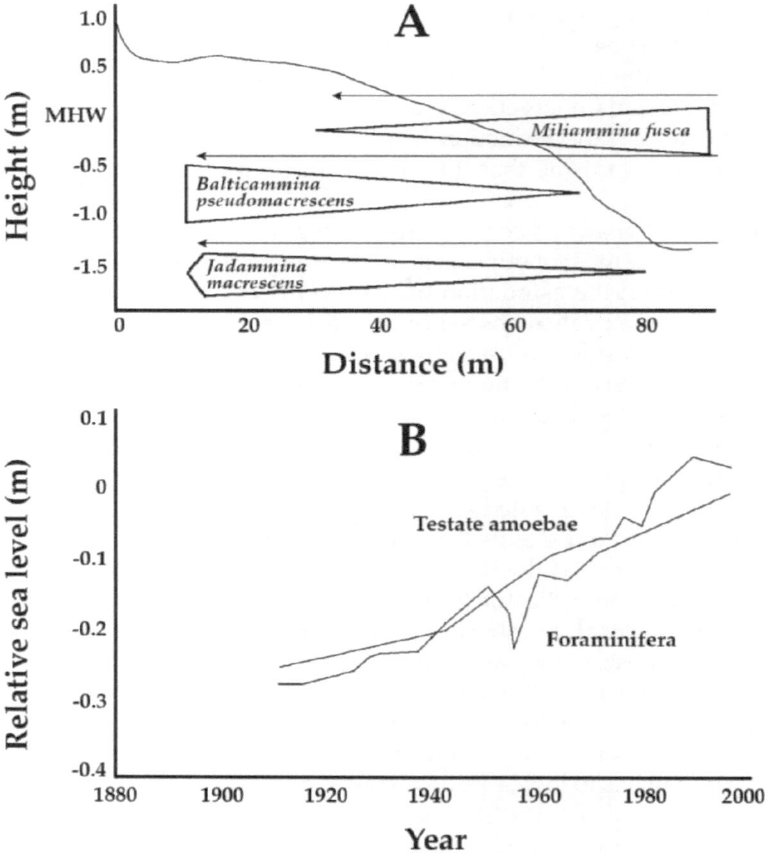

FIGURE 12.1

(A) The intertidal zonation of living and counterpart extinct foraminiferan species (shown by arrow) along the sloping surface considering of salt marsh in eastern Maine, as a representative example. (B) An example for sea-level rise in Nova Scotia, Canada based on salt-marsh testate amoebae and foraminifers (compiled, simplified and modified [for easier understanding] from Gehrels, 2013).

are one of the most persuasive environmental changes in the oceans (Feely et al., 2008). The progressive reduction in availability of carbonate ions (CO_3^{-2}) renders the acquisition of biogenic calcium carbonate ($CaCO_3$) by calcifying organisms more and more difficult and energetically costlier (Wood et al., 2008). The foraminifers, whose shell is mostly (90%) composed of calcium carbonate, may find it more expensive to acquire calcium carbonate (see p 17). However, no publication is yet available in this aspect, although a few publications are available on the effect of lake sediment acidity on the protozoan and their bacterivory (e.g. Tremaine and Mills, 1991).

Hyman (1940) indicated the most interesting 7-year long experiment by William Dallinger and John Drysdale (1874) on slow and gradual acclimation to elevated temperatures up to 70°C in three flagellate species. Unfortunately, their original publication/report is not available. However, their study established that flagellates can adapt as fast as the climate change elevating temperature. Experimental observations are reported for growth and reproduction over a thermal range of 10° to 25°C in a rhizopod *Acanthamoeba prolyphaga* and 8.5° to 20°C in a ciliate *Tetrahymena pyriformis* (Laybourn-Parry, 1987). In the marine parasitic ciliate *Zoothamnium hiketes* and *Anoplophyra filum*, adaptive changes occur within a day for elevated temperatures but within the natural biokinetic range (Ushakov, 1968). It is likely that some protozoans may adapt to increasing temperatures (e.g. *Zooxanthella*).

Testing the coccolithophore *Emiliania huxleyi* at 15°C (ambient) and 26.5°C in combination with pCO_2 400 (ambient), 1,100 (medium) and 2,200 (high) levels of pCO_2 concentrations, Schluter et al. (2014) reported that adaptation to temperature elevation occurred independent of acidification levels. Remarkably, growth rate increased up to 16% in the population exposed to one year long warming. Owing to adaptive evolution, the coccolithophore was 55 and 101% better than the non-adapted controls.

Symbiotic zooxanthellae in corals: Built over millions of years, coral reefs are the 'home' for > 25% of marine species. Therefore, they represent the biologically most diverse marine ecosystems. The very first global assessment revealed that a third of reef-building corals of the world are threatened with extinction potency. Approximately, 52% acroporid coral species (listed in the threatened list) encounter the higher risk level of extinction than the relatively more resistant Faviids and Porites (*sci.odu.edu/gmsa/about-/corals.shtml*). In the Gulf of Mannar (India), known for the abundance and diversity of corals, for example, coral bleaching is reported at least once during April–May every year. Prevalence of coral bleaching increased from 6.9 times during 2007 to 10.6 times during 2011 (see Pandian, 2021b). As reef-builders, the corals owe their success to symbiosis with the dinoflagellate zooxanthellae of the genus *Symbiodinium*. In coral tissue, algae live at densities of 10^6 cell/cm^2 and provide up to 90% of coral's nutrient requirements (Berkelmans and van Oppen, 2006) by releasing photosynthate in the form of glycerol (see Pandian, 1975). Feeding corals with *Artemia* larvae has shown that bleaching is caused by the cumulative effect of thermal stress and starvation. Perhaps, abundant prey availability for these "sit and watch" predators and faster feeding can save them from bleaching. For example, *Mytilus edulis*, a representative example for green shoots and new hopes, feeds more at a faster rate and meets the additional cost of acquiring calcium carbonate and avoids shell dissolution on exposure to acidic seawater (see Pandian, 2021b).

Experimental studies have shown that *Symbiodinium* within the coral is sensitive to increase in seawater surface temperatures (SST) of 24–28°C, whereas the coral can successfully tolerate up to 34°C and show no signs of apoptosis and necrosis even up to 36°C. Scleractinian corals are excellent biomarkers of thermal stress. Their optimal growth occurs between 26°C and 28°C but can exist between 18°C and 36°C (NOAA, 2009). Thus, the zooxanthellae are more sensitive to increase in SST, while corals are not. New hopes for saving corals come from more recent researches. 1. On exposure to SST stress, corals release a proportion of their zooxanthellae and reassociate with new combinations of zooxanthellae, that are better adapted to elevated temperature. However, the reassociation is limited to 10% only (Sammarco and Strychar, 2009). 2. In the symbiotic *Symbiodinium, in situ* clonal multiplication alone occurs. Hence, they have to depend only on random mutation to gain adaptive tolerance against thermal stress. Nevertheless, thanks to molecular geneticists (e.g. LaJeunesse and Trench, 2000, Stat et al., 2008), it was found that there is more than one species of *Symbiodinium*. Further, the family Symbiodiniaceae comprise several genera and each genus consist of many species, each of which includes a large number of clades and subclades (Hill et al., 2019). Therefore, *Symbiodinium* may have begun to witness the advent of new rapidly evolving clades that are responding to global warming.

Warner et al. (1996) brought interesting information on zooxanthella density and chlorophyll level as well as photosynthetic efficiency of zooxanthellae in four coral species. With increasing temperature from 28°C to 32°C, zooxanthella density decreased from 1.87–2.54 × 10^6 cell/cm^2 coral in *Agaricia lamarki* and *Siderastrea radius* to 2.13–0.58 × 10^6 cell/cm^2 in *Montastrea annularis* and *A. lamarki* (Table 12.2). Further, chlorophyll level in each living zooxanthella was also reduced from 1.85–4.40 pg/zooxanthella cell at 28°C in *S. radius* and *A. lamarki* to 2.03–3.19 pg/cell at 32°C in *M. annularis* and *A. lamarki*. Therefore, thermal stress not only decreased zooxanthella density but also chlorophyll level in each living zooxanthella cell (Table 12.2). As a consequence, photosynthetic efficiency of these zooxanthellae also decreased. Surprisingly, both the density of zooxanthellae and chlorophyll level of *Agaricia agaricites* increased from 1.04 × 10^6 cell/cm^2 at 28°C to 1.99 × 10^6 cell/cm^2 at 32°C and retained the same chlorophyll level of 2.14 pg/cell at these temperatures (Table 12.2). *Therefore, in one of four coral species, i.e. 25% of zooxanthellae are adapted to survive and photosynthesize more efficiently at increased temperature up to 32°C*. It is likely that *A. agaricites* has acquired one of those species or clades or sub-clades of *Symbiodinium* that are better adapted to tolerate and function efficiently at 32°C. *The dinoflagellates, some of which became symbionts in corals and are unable to tolerate temperatures above 32°C, appeared during mid-Jurassic* (Delwiche, 2007), *when the earth was already cooled*. Therefore, thermal tolerance of a specific plant taxon may have to be considered along with the thermal level, when it originated.

TABLE 12.2

Density of zooxanthellae and chlorophyll (after 48 hours isolation) as a function of temperature in some coral species (condensed from Warner et al., 1996). Arrows shows the direction of changes

Temperature (°C)	Zooxanthellae (no. × $10^6/cm^2$)	Chlorophyll (pg/cell)
	Siderastrea radius	
28	2.54 ↓	1.85 ↓
34	1.94	2.32
36	1.20	1.22
	Montastrea annularis	
28	2.23 ↓	2.88 ↓
32	2.13	2.03
34	1.34	1.12
	Agaricia lamarki	
28	1.87 ↓	4.40 ↓
32	0.58	3.19
	Agaricia agaricites	
24	1.26 ↑	1.63 ↑
28	1.04	2.15
32	1.99	2.14

With the appearance of more and more temperature-insensitive/resistant green shoots, which may flourish and diversify in the years to come, the mother earth shall continue to witness burgeoning life forms and their species diversity, hopefully inclusive of man.

13

References

Abe, T., Bignell, D.E. and Higashi, M. 2000. *Termites: Evolution, Sociality, Symbioses, Ecology*. Springer Science Business Media, p 469.

Adam, R.D. 2001. Biology of *Giardia lamblia*. Clinic Microbiol Rev, 14: 447–475.

Adlard, R.D. and O'Donoghue, P.J. 1998. Perspectives on the biodiversity of parasitic Protozoa in Australia. Int J Parasitol, 28: 887–897.

Agrawal, D.C. 2013. Average annual rainfall over the globe. Physics Teacher, 51: 540–541.

Ahmed, N.H. 2014. Cultivation of parasites. Trop Parasitol, 4: 80–89.

Andersen, P. and Fenchel, T. 1985. Bacterivory by microheterotrophic flagellates in seawater samples. Limnol Oceanogr, 30: 198–202.

Anderson, O.R. 1983. *Radiolaria*. Springer-Verlag, New York, p 355.

Anderson, O.R. 1988. Reproduction. In: *Comparative Protozoology*. Springer, Berlin, pp 375–392.

Anderson, O.R. 2001. Radiolarians. In: *Encyclopedia of Ocean Sciences*. Elsevier, Vol 4, pp 2315–2320.

Andreadis, T.G. 2007. Microsporidian parasites of mosquitoes. J Am Mosquito Control Ass, 23: 3–29.

Arnold, Z.M. 1966. Observations on the sexual generation of *Gromia oviformis* Dujardin. J Protozool, 13: 23–27.

Aufderheide, K.J., Frankel, J. and Williams, N.E. 1980. Formation and positioning of surface-related structures in Protozoa. Microbiol Rev, 44 : 252–302.

Balamuth, W. 1940. Regeneration in Protozoa: A problem of morphogenesis. Q Rev Biol, 14: 290–337.

Baniya, C.B., Solhoy, T., Gauslaa, Y. and Palmer, M.W. 2010. The elevation gradient of lichen species richness in Nepal. Lichenol, 42: 83–96.

Bardele, C.F. 1970. Budding and metamorphosis in *Acineta tuberosa*. An electron microscopic study on morphogenesis in Suctoria. J Protozool, 17: 51–70.

Baxter, J.M. and Jones, A.M. 1978. Growth and population structure of *Lepidochitona cinereus* (Mollusca: Polyplacophora) infected by *Murichinia chitonis* (Protozoa: Sporozoa) at Easthaven, Scotland. Mar Biol, 46: 305–313.

Bayless, B.A., Navarro, F.M. and Winey, M. 2019. Motile cilia: Innovation and insight from ciliate model organism. Cell Dev Biol, 7, DOI: 10.3389/fcell.2019.00265.

Bekker, A., Holland, H.D., Wang, P.-L. et al. 2004. Dating the rise of atmospheric oxygen. Nature, 427: 117–120.

Belzecki, G., Miltko, R., Michalowski, T. and McEwan, N.R. 2015. Methods for the cultivation of ciliated protozoa from the large intestine of horses. FEMS Microbiol Lett, 363, DOI: 10.1093/femsle/fnv233.

Berkelmans, R. and van Oppen, M.J.H. 2006. The role of zooxanthellae in the thermal tolerance of corals: a 'nugget of hope' for coral reefs in an era of climate change. Proc R Soc, 273B: 2305–2312.

Biddick, C.J., Rogers, L.H. and Brown, T.J. 1984. Viability of pathogenic and non-pathogenic free-living amoebae in long-term storage at a range of temperatures. Appl Env Microbiol, 48: 859–860.

Boenigk, J. and Novarino, G. 2004. Effect of suspended clay on the feeding and growth of bacterivorous flagellates and ciliates. Aquat Microbiol Ecol, 34: 181–192.

Borradaile, L.A., Potts, F.A. Eastham, L.E. et al. 1977. *The Invertebrata*. Revised by Kerkut, G.A., Cambridge University Press, p 820.

Bowers, E.A. 1969. *Cercaria bucephalopsis haimeana* (Lacaze-Dutheires) (Digenea: Bucpehalidae) in the cockle *Cardium edule* L. in South Wales. J Nat Hist, 3: 409–422.

Brown, E.M. 1963. Studies on *Cryptocaryon irritans* Brown. Progress in Protozoology. Academia Publ House, Prague, pp 583–607.

Butlin, R. 2002. Evolution of sex: The costs and benefits of sex: New insights from old asexual lineages. Nat Rev Genet, 3: 311–317.

Cali, A., Bencel, J.J. and Takvorian, P.M. 2017. Microsporidia. In: *Handbook of Protists* (eds). Archibald, J.M. et al., Springer International, pp 1559–1618.

Calvin, M. 1964. The path of carbon in photosynthesis, VI. J Chem Edu, 26(12), DOI; 10.1021/ed026p639.

Campbell, A.S. 1926. The cytology of *Tintinnopsis nucula* (Fol) Laackmann, with an account of its neuromotor apparatus, division, and with a description of a new intranuclear parasite. Univ Calif Publ Zool, 29: 179–236.

Capriulo, G.M. 1990. *Ecology of Marine Protozoa*. Oxford University Press, p 384.

Carrias, J.-F. Amblard, C. and Bourdier, G. 1996. Protistan bacterivory in an oligomesotrophic lake: Importance of attached ciliates and flagellates. Microb Ecol, 31: 249–268.

Carvalho, A.B. 2003. The advantages of recombination. Nat Genet, 32: 128–129.

Charman, D.J., Gehrels, W.R., Manning, C. and Sharma, C. 2010. Reconstruction of recent sea-level change using testate amoebae. Quarterner Res, 73: 208–291.

Coelho, S.M., Mignerot, L. and Cock, J.M. 2019. Origin and evolution of sex-determination systems in the brown algae. New Phytol, 222: 1751–1756.

Corliss, J.O. 1977. Annotated assignment of families and genera to the orders and classes currently comprising the corlissian scheme of higher classification for the phylum Ciliophora. Trans Am Microsc Soc, 96: 104–140.

Corliss, J.O. 1979. *The Ciliated Protozoa*. Pergamon Press, London, p 472.

Corliss, J.O. 2001. Protozoan taxonomy and systematics. In: *Encyclopedia of Life Sciences*. Wiley & Sons, London, pp 1–7.

Corliss, J.O. and Esser, S.C. 1974. Comments on the role of the cyst in the life cycle and survival of free-living protozoa. Trans Am Microsc Soc, 93: 578–593.

Costello, M.J. and Chaudhary, C. 2017. Marine biodiversity, biogeography, deep-sea gradients, and conservation. Curr Biol, 27R: 11–27.

Cox, F.E.G. 2002. Systematics of the parasitic Protozoa. Trends Parasitol, 18: 108.

Cuvier, G. 1816. *Le regne animal distribute d' apre's son organization*. Chez Deterville, Paris, p 252.

d'Avila-Levy, C.M., Boucinha, C., Kostygov, A. et al. 2015. Exploring the environmental diversity of kinetoplastid flagellates in the high-throughput DNA sequencing era. Mem Inst Oswaldo Cruz, 110: 956–965.

Dalton, C., Goater, A.D., Pethig, R. and Smith, H.V. 2001. Viability of *Giardia intestinalis* cysts and viability and sporulation state of *Cyclospora cayetanensis* oocysts determined by electrorotation. Appl Env Microbiol, 67: 586–590.

Damborenea, C. 2020. Flagellated Protozoa. In: *Thorp and Covich's Freshwater Invertebrates: Keys to Neotropical and Atlantic Fauna*. Volume 5. (eds). Rogers, D.C. and Thorp, J.H., Academic Press, pp 74–78.

Decelle, J., Martin, P., Paborstava, K. et al. 2013. Diversity, ecology and biogeochemistry of cyst-forming *Acantharia* (Radiolaria) in the oceans. PLoS One, 8: 353598.

Dehority, B.A. 2002. Gastrointestinal tracts of herbivores, particularly the ruminant: Anatomy, physiology and microbial digestion of plants. J Appl Anim Res, 21: 145–160.

DeLaca, T.E. 1982. Use of dissolved amino acids by the foraminifer *Notodendrodes antarctikos*. Am Zool, 22: 683–690.

Delwiche, C.F. 2007. The origin and evolution of dinoflagellates. In: *Evolution of Primary Producers in the Sea*. Academic Press, USA, pp 191–205.

Didier, E.S. and Weiss, L.M. 2006. Microsporidiosis: current status. Curr Opin Infec Dis, 19: 485–492.

Doerder, F.P. 2014. Abandoning sex: multiple origins of asexuality in the ciliate *Tetrahymena*. BMC Evol Biol, 14: 112, https://www.biomedcentral.com/1471-2148/14/112.

Eloff, A.K. and Hoven, W.v. 1980. Intestinal protozoa of the African elephant *Loxodonta africana* (Blumenbach). S Afri Tydskr Dierk, 15: 83–89.

Eikrem, W., Medlin, L.K., Henderiks, J. et al. 2017. Haptophyta. In: *Handbook of the Protists* (eds). Archibald J.M. et al., Springer, Switzerland, pp 1–61.
El-Bawab, F. 2020. Phylum Protozoa. In: *Invertebrate Embryology and Reproduction* (ed). El-Bawab, F., Academic Press, pp 68–102
Endo, Y., Fujii, D., Nishitani, G. and Wiebe, P.H. 2017. Life cycle of the suctorian ciliate *Ephelota plana* attached to the krill *Euphausia pacifica*. J Exp Mar Biol Ecol, 486: 368–372.
Engel, P. and Moran, N.A. 2013. The gut microbiota of insects—diversity in structure and function. Microbiol Rev, 37: 699–735.
Farley, C.A. 1967. A proposed life cycle of *Minchinia nelsoni* (Haplosporida, Haplosporiidae) in American oyster *Crassostrea virginica*. J Protozool, 14: 616–625.
Feely, R.A., Sabine, C.L., Hernandez-Ayon, J.M. et al. 2008. Evidence for upwelling of corrosive acidified water onto the continental shelf. Science, 320: 1490–1492.
Fenchel, T. 1980a. Suspension feeding in ciliated protozoa: structure and function of feeding organelles. Arch Protistenk, 123: 239–260.
Fenchel, T. 1980b. Suspension feeding in ciliated protozoa: feeding rates and their ecological significance. Microbiol Ecol, 6: 13–25.
Fenchel, T. 1986. Protozoan filter feeding. Prog Protistol, 1: 65–113.
Fenchel, T. and Finlay, B.J. 1983. Respiration rates in heterotrophic, free-living Protozoa. Microbiol Ecol, 9: 99–122.
Fenchel, T. and Finlay, B.J. 1995. *Ecology and Evolution in Anoxic Worlds* (eds). May, R.M. and Harvey, P.H., Oxford University Press, p 288.
Fenchel, T. and Finlay, B.J. 2006. The diversity of microbes: Resurgence of the phenotype. Phil Trans R Soc, 361B: 1965–1973.
Fenchel, T. and Harrison, P. 1976. The significance of bacteria grazing and mineral cycling for the decomposition of particulate detritus. In: *The Role of Terrestrial and Aquatic Organisms in Decomposition Processes* (eds). Anderson, J.M. and McFayden, A., Oxford, Blackwell, pp 285–299.
Ferguson, J.C. 1971. Uptake and release of free amino acids by star fishes. Biol Bull, 141: 22–29.
Ferguson, J.C. 1972. A comparative study of the net metabolic benefits derived from the uptake and release of free amino acids by marine invertebrates. Biol Bull, 162: 1–7.
Finlay, B.J. and Esteban, G.F. 1998. Freshwater protozoa: biodiversity and ecological function. Biodiv Conserv, 7: 1163–1186.
Finlay, B.J. and Esteban, G.F. 2018. Protozoa. In: *Reference Module in Life Sciences* (ed). Roitberg, B.D., Elsevier, pp 1–12.
Firkins, J.L., Yu, Z., Park, T. and Plank, J.E. 2020. Extending Burk Dehority's perspectives on the role of ciliate protozoa in the rumen. Front Microbiol, 11: 123.
Flowers, J., Li, S.I., Stathos, A. et al. 2010. Variation, sex, and social cooperation: Molecular population genetics of the social amoeba *Dictyostelium discoideum*. PLoS Genetics, 6: 31001013.
Foissner, W. 2014. Protozoa. In: *Reference Module in Earth Systems and Environmental Sciences*. Elsevier, pp 1–12.
Fuente, G.D.L., Skirnisson, K. and Dehority, B.A. 2006. Rumen ciliate fauna of Icelandic cattle, sheep, goats and reindeer. Zootaxa, 1377: 47–60.
Gaines, A., Ludovice, M., Xu, J. et al. 2019. The dialogue between protozoa and bacteria in a microfluidic device. PLoS One, 14: e0222484.
Gardiner, M.S. 1972. *The Biology of Invertebrates*. McGraw-Hill Book Company, London, p 978.
Gehrels, W.R. 2002. Intertidal foraminifera as palaeoenvironmental indicators. In: *Quarternary Environmental Micropalaeontology* (ed). Haslett, S.K., Arnold Publishers, London, pp 91–114.
Gehrels, W.R. 2013. Microfossil-based reconstructions of Holocene relative sea-level change. http://dx.doi.org/10.1016/B978-0-444-53643-3.00137-0.
Gibson, B., Wilson, D.J., Feil, E. and Eyre-Walker, A. 2018. The distribution of bacterial doubling times in the wild. Proc R Soc, 285B: 20180789.
Gijzen, H.J. and Barughare, M. 1992. Contribution of anaerobic Protozoa and methanogens to hindgut metabolic activities of the American cockroach, *Periplanata americana*. Appl Env Microbiol, 58: 2565–2570.

Ginger, M.L., Portman, N. and McKean, P.G. 2008. Swimming with protists: Perception, motility and flagellum assembly. Nat Rev Microbiol, 6: 838–850.
Gnanamuthu, C.P. 1943. The Foraminifera of Krusadai Island in the Gulf of Mannar. Bull Mad Govt Mus New Ser, 5: 1–21.
Goldstein, S.T. 1999. Foraminifera: A biological overview. In: *Modern Foraminifera* (ed). Sen Gupta, B.K.S. Kluwer Academic Publishers, UK, pp 37–55.
Grell, K.G. 1973. *Protozoology*. Toppan Printing, Singapore, p 544.
Grujcic, V., Kasalicky, V. and Simek, K. 2015. Prey-specific growth responses of freshwater flagellate communities induced by morphologically distinct bacteria from the genus *Limnohabitans*. Appl Environ Microbiol, 81: 4993–5002.
Guiry, M.D. 2012. How many species of algae are there? J Phycol, 48: 1057–1063.
Gupta, S. and Das, S.R. 1999. Stock cultures of free-living Amebas: Effect of temperature on viability and pathogenicity. J Parasitol, 85: 137–139.
Hallock, P. 1999. Symbiont-bearing Foraminifera. In: *Modern Foraminifera* (ed). Gupta, B.K.S. Kluwer Academic Publisher, UK, pp 123–139.
Haq, B.U. and Al-Qahtani, A.M. 2005. Phanerozoic cycles of sea-level change on the Arabian platform. GeoArabia, 10: 127–160.
Harris, E., Detmer, J., Dungan, J. et al. 1996. Detection of *Trypanosoma brucei* spp. in human blood by a nonradioactive branched DNA-based technique. J Clinic Microbiol, 34: 2401–2407.
Hawes, R.S.J. 1963. The emergence of asexuality in Protozoa. Q Rev Biol, 38: 232–242.
Hawksworth, D.L. and Lucking, R. 2017. Fungal diversity revisited: 2.2 to 3.8 million species. Microbiol Spectrum, 5: FUNK-0052-2016.
Heinz, P., Geslin, E. and Hemleben, C. 2005. Laboratory observations of benthic foraminiferal cysts. Mar Biol Res, 1: 149–159.
Hill, L.J., Paradas, W.C., Willemes, M.J. et al. 2019. Acidification-induced cellular changes in *Symbiodinium* isolated from *Mussismilia braziliensis*. PLoS One, 14: e0220130.
Hochberg, F.G. 1983. The parasites of cephalopods: A review. Mem Nah Mus Vic, 44: 109–145.
Hochberg, F.G. 1990. Diseases caused by Protisteans and Metazoans. In: *Diseases of Marine Animals* (ed). Kinne, O., Biologische Anstalt Helgoland, Hamburg, Vol 3, pp 47–200.
Holt, H.R., Selby, R., Mumba, C. et al. 2016. Assessment of animal African trypanosomiasis (AAT) vulnerability in cattle-owning communities of sub-Saharan Africa. Parasites Vectors, 9: 53.
Horan, N. 2003. Protozoa. In: *The Handbook of Water and Wastewater Microbiology* (eds). Mara, D. and Horan, N. Academic Press, pp 69–76.
Hyman, L.H. 1940. Protozoa. In: *The Invertebrates: Protozoa through Ctenophora*. McGraw-Hill Book, New York, pp 44–232.
Ismael, A.A. 2003. Succession of heterotrophic and mixotrophic dinoflagellates as well as autotrophic microplankton in the harbour of Alexandria, Egypt. J Plankton Res, 25: 193–202.
IPCC. 2021. Climate Change 2021: The Physical Science Basis. https://www.ipcc.ch/report/ar6/wg1/.
Jahn, T.L. 1934. Problems of population growth in the Protozoa. Cold Spring Harb Symp Quant Biol, 2: 167–180.
Jalovecka, M., Hajdusek, O., Sojka, D. et al. 2018. The complexity of piroplasms life cycles. Front Cell Infect Microbiol, 8: 248, DOI: 10.3389/fcimb.2016.00248.
Jangoux, M. 1990. Diseases in Echinoderms. In: *Diseases of Marine Animals* (ed). O. Kinne. Biologische Anstalt Helgoland, Hamburg, Vol 3, Part 2, pp 439–568.
Jennings, H.S. 1939. Genetics of *Paramecium bursaria*. I. Mating types and groups, their interrelations and distribution; mating behavior and self sterility. Genetics, 24: 202–233.
Jennings, J.B. 1997. Nutritional and respiratory pathways to parasitism exemplified in the Turbellaria. Int J Parasitol, 27: 679–691.
Jiang, C-Q, Wang, G-Y, Yang, W-T. et al. 2018. Insights into the origin and evolution of Peritricha (Oligohymanophorea, Ciliophora) based on analyses of morphology and phylogenomics. Mol Phylogenet Evol, DOI: https://doi.org/10.1016/j.ympev.2018.11.018.
Jorgensen, C.B. 1966. *Biology of Suspension Feeding*. Pergamon Press, p 372.

Karpov, S.A. 2000. Flagellate phylogeny: An ultrastructural approach. In: *Flagellates: Unity, Diversity and Evolution* (eds). Leadbeater, B.S.C. and Green, J.C., Taylor and Francis, London, pp 336 – 360.
Karpov, S.A. 2016. Flagellar apparatus structure of choanoflagellates. Cilia, 5: 11.
Kelley, I. and Pfiester, L.A. 1990. Sexual reproduction in the freshwater dinoflagellate *Gloeodinium montanum*. J Phycol, 26: 167–175.
Khan, R.A. 1980. The leech as a vector of a fish piroplasm. Can J Zool, 58: 1631–1637.
Khanna, D.R. 2004. *Biology of Protozoa*. Discovery Publishing, Delhi, p 366.
King, N., Westbrook, M.J., Young, S.L. et al. 2008. The genome of the choanoflagellate *Monosiga brevicollis* and the origin of metazoans. Nature, 451: 783–788.
King, N., Young, S.L., Abedin, M. et al. 2009. Starting and maintaining *Monosiga brevicollis* cultures. Cold Spring Harb Protoc, 2009: pdb.prot5148.
Kinne, O. 1983a. *Diseases of Marine Animals*. Biologisch Anstalt Helgoland, Hamburg, Volume 2, Part 1, p 1038.
Kinne, O. 1983b. Introduction. In: *Diseases of Marine Animals*. Biologisch Anstalt Helgoland, Hamburg, Volume 3, Part 1, pp 1–20.
Kinne, O. 1984. *Diseases of Marine Animals*. Biologisch Anstalt Helgoland, Hamburg, Volume 4, Part 1, p 541.
Kinne, O. 1990. *Diseases of Marine Animals*. Biologisch Anstalt Helgoland, Hamburg, Volume 3, Part 2, p 696.
Klein, S.L. 2004. Hormonal and immunological mechanisms mediating sex differences in parasitic infection. Parasite Immunol, 26: 247–264.
Kosakyan, A., Siemensma, F., Fernandez, L.D. et al. 2020. Amoebae. In: *Thorp and Covich's Freshwater Invertebrates: Keys to Neotropical and Atlantic Fauna*. Volume 5. (eds). Rogers, D.C. and Thorp, J.H., Academic Press, pp 13–36.
Krebs, H.A. 1940. The citric acid cycle. Biochem J, 34: 460–463.
Kudo, R. 1933. A taxonomic consideration of Myxosporidea. Trans Am Microsc Soc, 52: 195–216.
Kuppers, G.C., Claps, M.C. and Paiva, T. da S. 2020. Ciliphora. In: *Thorp and Covich's Freshwater Invertebrates: Keys to Neotropical and Atlantic Fauna*. Volume 5. (eds). Rogers, D.C. and Thorp, J.H., Academic Press, pp 37–73.
Kuroiwa, H., Nozaki, H. and Kuroiwa, T. 1993. Preferential digestion of chloroplast nuclei in sperms before and during fertilization in *Volvox carteri*. Cytologia, 58: 281–291.
Lahr, D.J.G., Grant, J., Nguyen, T. et al. 2011. Comprehensive phylogenetic reconstruction of Amoebozoa based on concatenated analyses of SSU-rDNA and Actin genes. PLoS One, 6: e 22780.
LaJeunesse, T.C. and Trench, R.K. 2000. Biogeography of two species of *Symbiodinium* (Freudenthal) inhabiting the intertidal sea anemone *Anthopleura elegantissima* (Brandt). Biol Bull, 199: 126–134.
Larson, B.T., Ruiz-Herrero, T., Lee, S. et al. 2020. Biophysical principles of choanoflagellate self-organization. Proc Natl Acad Sci USA, 117: 1303–1311.
Lauckner, G. 1983. Diseases of Mollusca: Bivalvia. In: *Diseases of Marine Animals* (ed). O. Kinne. Biologische Anstalt Helgoland, Hamburg, Vol 2, pp 477–962.
Laybourn, J. and Stewart, J.M. 1975. Studies on consumption and growth in the ciliate *Colpidium campylum* Stokes. J Anim Ecol, 44: 165–174.
Laybourn, J.E.M. and Finlay, B.J. 1976. Respiratory energy losses related to cell weight and temperature in ciliated protozoa. Oecologia, 24: 349–355.
Laybourn-Parry, J. 1984. Physiological functioning of Protozoa. In: *A Functional Biology of Free-Living Protozoa*. Springer, Boston, pp 66–109.
Laybourn-Parry, J. 1987. Protozoa. In: *Animal Energetics* (eds). Pandian, T.J. and Vernberg, F.J. Academic Press, San Diego, Vol 1, pp 1–26.
Leadbeater, B.S.C. and Karpov, S. 2000. Cyst formation in a freshwater strain of the choanoflagellate *Desmarella moniliformis* Kent. J Eukaryot Microbiol, 47: 433–439.
Lee, B.Y., Bacon, K.M., Bottazzi, M.E. and Hotez, P.J. 2013. Global economic burden of Chagas disease: a computational simulation model. Lancet Infect Dis, 13: 342–348.
Lee, J.J. 2006. Algal symbiosis in larger foraminifera. Symbiosis, 42: 3–75.

Lee, W.J. and Patterson D.J. 1998. Diversity and geographic distribution of free-living heterotrophic flagellates – analysis of PRIMER. Protist, 149: 229–243.

Leger, L. 1909. Les Schizogregarines des Tracheates. II. Le genre *Schizocystis* Arch Protistenk, 18: 83–110.

Lekfeldt, J.D.S. and Ronn, R. 2008. A common soil-flagellate (*Cercomonas* sp.) grows slowly when feeding on the bacterium *Rhodococcus fascians* in isolation, but does not discriminate against it in a mixed culture with *Sphingopyxis witflariensis*. FEMS Microbiol Ecol, 65: 113–124.

Levin, T.C. and King, N. 2013. Evidence of sex and recombination in the Choanoflagellate *Salpingoeca rosetta*. Curr Biol, 23: 2176–2180.

Levine, N.D. 1980. Some corrections of coccidian (Apicomplexa: Protozoa) nomenclature. J Parasitol, 66: 830–832.

Levine, N.D., Corliss, J.O., Cox, F.E.G. et al. 1980. A newly revised classification of the Protozoa. J Protozool, 27: 37–58.

Lewis, J., Harris, A.S.D. and Jones, K.J. 1999. Long-term survival of marine planktonic diatoms and dinoflagellates in stored sediment samples. J Plankton Res, 21: 343–354.

Lewis, W.M. 1983. Interruption of synthesis as a cost of sex in small organisms. Am Nat, 121: 825–834.

Lodish, H., Berk, A., Zipursky, S.L. et al 2000. *Molecular Cell Biology*. W.H. Freeman, New York, p 1280.

Lom, J. 1984. Diseases caused by Protisteans. In: *Diseases of Marine Animals* (ed). Kinne, O., Biologische Anstalt Helgoland, Hamburg, Vol 4, Part 1, pp 114–168.

Long, R.L., Gorecki, M.J., Renton, M. et al. 2015. The ecophysiology of seed persistence: A mechanistic view of the journey to germination or demise. Biol Rev, 90: 31–59.

Lowe, C.D., Day, A., Kemp, S.J. and Montagnes, D.J.S. 2005. There are high levels of functional and genetic diversity in *Oxyrrhis marina*. J Euk Microbiol, 52: 250–257.

Lucchesi, P. and Santangelo, G. 2004. How often does conjugation in ciliates occur? Clues from a seven-year study on marine sandy shores. Aquat Microbiol Ecol, 36: 195–200.

Lundholm, N., Ribeiro, S., Andersen, T.J. et al. 2011. Buried alive – germination of up to century-old protist resting stages. Phycologia, 50: 629–640.

Lynn, D.H. 2010a. Introduction and progress in the last half century. In: *The Ciliated Protozoa*. Springer, Dordrecht, pp 1–14.

Lynn, D.H. 2010b. Phylum Ciliophora – conjugating, ciliated protists with nuclear dualism. In: *The Ciliated Protozoa*. Springer, Dordrecht, pp 89–120.

Mayen-Estrada, R. and Utz, L.R.P. 2018. A checklist of species of Vorticellidae (Ciliophora: Peritricha) epibionts of crustaceans. Zootaxa, 4500: 301–328.

Mayr, E. 1942. *Systematics and the Origin of Species*. Columbia University Press, New York, p 334.

Mazur, T., Hadas, E. and Iwanicka, I. 1995. The duration of the cyst stage and the viability and virulence of *Acanthamoeba* isolates. Trop Med Parasitol, 46: 106–108.

Mehlhorn, H. 2008. Opalinata. In: *Encyclopedia of Parasitology*. 3rd Edition, Springer, Berlin pp 1049–1050.

Mehlhorn, H. 2016. *Encyclopedia of Parasitology*. 4th Edition, Springer, Berlin p 3084.

Meibalan, E. and Marti, M. 2017. Biology of Malaria transmission. In: *Additional Perspectives on Malaria: Biology in the Era of Eradication* (eds). Wirth, D.F. and Alonso, P.L., Cold Spring Harbor Laboratory Press, pp 27–41.

Menezes, M. and Bicudo, C.E. de M. 2010. Freshwater Raphidophyceae from the State of Rio de Janeiro, Southwest Brazil. Biota Neotrop, 10: 323–331.

Meyers, T.R. 1990. Diseases caused by Protisteans and Metazoans. In: *Diseases of Marine Animals* (ed). Kinne, O., Biologische Anstalt Helgoland, Hamburg, Vol 3, Part 2, pp 350–424.

Michaiowski, T. 2005. Rumen protozoa in the growing domestic ruminant. In: *Biology of Growing Animals: Microbial Ecology in Growing Animals* (eds). Holzapfel, W.H., Naughton, F.J., Pierzynowski, S.G. et al., Elsevier Health Science, Vol 2, pp 54–74.

Mikhalevich, V.I. 2021. The current state of foraminiferal research and future goals. Int J Paleobot Paleontol, 4: 000125.

Miller, N.G., Wassenaar, L.I., Hobson, K.A. and Norris, D.R. 2012. Migratory connectivity of the Monarch butterfly (*Danaus plexippus*): patterns of spring re-colonization in Eastern North America. PLoS One, 7: e31891.

Mikhalevich, V.I. 2021. The current state of foraminiferal research and future goals. Int J Paleobiol Paleontol, 4: 000125.
Monniot, C. 1990. Diseases of Urochodata. In: *Diseases of Marine Animals* (ed). Kinne, O., Biologische Anstalt Helgoland, Hamburg, Vol 3, Part 2, pp 569–625.
Murray, J.W. 2007. Biodiversity of living benthic foraminifera: How many species are there? Mar Micropaleontol, 64: 163–176.
Murugesan, P., Punniyamoorthy, R. and Mahadevan, G. 2021. *Monograph on benthic Foraminifera of Tamil Nadu coastal waters, Southeast India*. Ministry of Earth Science, Government of India, p 240.
Naiyer, S., Bhattacharya, A. and Bhttacharya, S. 2019. Advances in *Entamoeba histolytica* biology through transcriptomic analysis. Front Microbiol, 10: 1921, DOI: 01.3389/fmicb.2019.01921.
National Oceanic and Atmospheric Administration (NOAA). 2009. Coral health and monitoring program. Nat Oceanic Atmosphere, Silver Springs, USA, *https://www.coral.noaa.gov/faq1.shtml.*
Newbold, C.J., Fuente, G.de.la., Belanche, A. et al. 2015. The role of ciliate Protozoa in the rumen. Front Microbiol, 6: 1313, DOI: 10.3389/fmicb.2015.01313.
Nielsen, L.T., Aadzadeh, S.S., Dolger, J. et al. 2017. Hydrodynamics of microbial filter feeding. Proc Natl Acad Sci USA, 114: 9373–9378.
Noble, E.R. 1944. Life cycles in the Myxosporidea. Q Rev Biol, 19: 213–235.
Noda, S., Hongoh, Y., Sato, T. and Ohkuma, M. 2009. Complex coevolutionary history of symbiotic bacteroidales bacteria of various protists in the gut of termites. BMC Evol Biol, 9: 158.
Ohkuma, M. 2003. Termite symbiotic systems: Efficient bio-recycling of lignocellulose. Appl Microbiol Biotechnol, 61: 1–9.
Ohkuma, M. 2008. Symbioses of flagellates and prokaryotes in the gut of lower termites. Trends Microbiol, 16: 345–352.
Ohkuma, M., Noda, S., Hongoh, Y. et al. 2009. Inheritance and diversification of symbiotic trichonymphid flagellates from a common ancestor of termites and the cockroach *Cryptocercus*. Proc R Soc, 276B: 239–245.
Okada, H. and Honjo, S. 1973. The distribution of oceanic coccolithophores in the Pacific. Deep-Sea Res Oceanogr Abst, 20: 355–374.
Overstreet, R.M. 1978. Marine Maladies? Worms, germs and other symbionts from Northern Gulf of Mexico. MASGP-78-021, Mississippi-Alabama Sea Grant Consortium. Ocean Springs, Mississippi.
Overstreet, R.M. and Weidner, E. 1974. Differentiation of microsporidian spore tails in *Indosporus spraguie* gen. et sp. n. z. Parasikde, 44: 169–186.
Pablos Torro, L.M. de. and Morales, J-L. 2018. *Protozoan Parasitism: From Omics to Prevention and Control*. Caister Academic Press, p 182.
Paget, T.A. and Lloyd, D. 1990. Trichomonas vaginalis requires traces of oxygen and high concentrations of carbon dioxide for optimal growth. Mol Biochem Parasitol, 41: 65–72.
Pandey, S.N. and Trivedi, P.S. 1995. *A Textbook of Algae*. Vikas Publishing House, New Delhi, p 342.
Pandian, T.J. 1975. Mechanism of Heterotrophy. In: *Marine Ecology* (ed). Kinne, O., John Wiley, London, 3A: 61–249.
Pandian, T.J. 1994. Crustacea. In: *Reproductive Biology of Invertebrates* (eds). Adiyodi, K.G. and Adiyodi, R.G. Oxford & IBH Publishing, New Delhi, 6A: 39–166.
Pandian, T.J. 2011. *Sexuality in Fishes*. Science Publishers/CRC Press, USA, p 218.
Pandian, T.J. 2015. *Environmental Sex Determination in Fish*. CRC Press, USA, p 299.
Pandian, T.J. 2016. *Reproduction and Development in Crustacea*. CRC Press, USA, p 301.
Pandian, T.J. 2017. *Reproduction and Development in Mollusca*. CRC Press, USA, p 299.
Pandian, T.J. 2018. *Reproduction and Development in Echinodermata and Prochordata*. CRC Press, US, p 270.
Pandian, T.J. 2019. *Reproduction and Development in Annelida*. CRC Press, USA, p 276.
Pandian, T.J. 2020. *Reproduction and Development in Platyhelminthes*. CRC Press, USA, p 303.
Pandian, T.J. 2021b. *Evolution and Speciation in Animals*. CRC Press, p 346.
Pandian, T.J. 2022. *Evolution and Speciation in Plants*. CRC Press, p 354.

Patterson, D.J. 1998. *Free-living Freshwater Protozoa. A Ecological Guide*. John Wiley, New York, p 223.

Pawestri, W., Nurairi, D.M. and Andityas, M. 2019. The estimation of economic losses due to coccidiosis in broiler chickens in Central Java, Indonesia. IPO Conf Series: Earth and Environmental Sciences, 411: 012030.

Peacock, L., Bailey, M., Carrington, M. and Gibson, W. 2014. Meiosis and haploid gametes in the pathogen *Trypanosoma brucei*. Curr Biol, 24: 181–186.

Pestel, B. 1931. Beitrage zur Morphologie und Biologie des *Dendrocometes paradoxus* Stein. Arch Protistenk, 75: 403–471.

Poljansky, G.I. 1992. Protozoology and the problem of species. J Protozool, 39: 177–180.

Probert, I., Siano, R., Poirier, C. et al. 2014. *Brandtodinium* Gen. Nov. and *B. nutricula* Comb. Nov. (Dinophyceae), a dinoflagellate commonly found in symbiosis with polycystine radiolarians. J Phycol, 50: 388–399.

Prytherci, H.F. 1940. The life cycle and morphology of *Nematopsis ostrearum* sp. nov., a gregarine parasite of the mud crab and oyster. J Morph, 66: 39–65.

Putter, A. 1909. *Die Einbrung der Wassertiere und der Stoffhaushalt der Gewasser*. G Fischer, Jena.

Randall, J.T. and Hopkins, J.M. 1962. On the stalks of certain peritrichs. Phil Trans R Soc, 245B: 59–79.

Ricci, N., Verni, F. and Rosati, G. 1985. The cyst of *Oxytricha bifaria* (Ciliata: Hypotrichida). I. Morphology and significance. Trans Am Microsc Soc, 104: 70–78.

Robert-Gangneux, F. and Darde, M.L. 2012. Epidemiology and diagnostic strategies for Toxoplasmosis. Clinic Microbiol Rev, 25: 264–296.

Rosati, G., Verni, F. and Nieri, L. 1983. Investigation on the cyst wall of the hypotrich ciliate *Gastrostyla steini* Engelmann. Monit Zool Ital, 17: 19–26.

Rosenbaum, J.L. and Child, F.M. 1967. Flagellar regeneration in protozoan flagellates. J Cell Biol, 34: 345–364.

Ross, B.J. and Hallock, P. 2016. Dormancy in the Foraminifera: A review. J Foraminiferal Res, 46: 358–368.

Salmaso, N. and Tolotti, M. 2009. Other phytoflagellates and groups of lesser importance. In: *Encyclopedia of Inland Waters* (ed). Likens, G.E., Academic Press, pp 174–183.

Sammarco, P.W. and Strychar, K.B. 2009. Effects of climate change/global warming on coral reefs: adaption/exaptation in corals, evolution in Zooxanthellae, and biogeographic shifts. Environ Bioindic, 4: 9–35.

Sanders, R.W. 2009. Protists. In: *Encyclopedia of Inland Waters*. Academic Press, Amsterdam, pp 252–260.

Schlegel, M. and Meisterfeld, R. 2003. The species problem in protozoa revisited. Eur J Protistol, 39: 349–355.

Schluter, L., Lobeck, K.T., Gutowska, M.A. et al. 2014. Adaptation of a globally important coccolithophore to ocean warming and acidification. Nature Climate Change, 4, www. nature/com/natureclimatechange

Schluter, L., Lohbeck. K.T., Gutowska, M.A. et al. 2014. Adaptation of a globally important coccolithophore to ocean warming and acidification. Nature Climate Change, 4: 1024–1030.

Schmiedl, G. 2019. *Use of Foraminifera in Climate Science*. Oxford Research Encyclopaedia of climate Science, Oxford University Press, DOI: 10.1093/acrefore/9780190228620.013.735.

Schrevel, J. 1969. Recherches sur le cycle des Lecudinidae gregarines parasites d' annelids polychetes. Protistologica, 5: 561–588.

Schrevel, J. 1971. Contribution a l'etude a l'etude des Selenidiidae parasites d'Annelides Polychetes: II. Ultrastructure de quelques trophozoites. Protistologica, 7: 101–130.

Schrevel, J. and Desportes, I. 2016. Gregarines. In: *Encyclopedia of Parasitology* (ed). Mehlhorn, H. Springer, pp 1142–1188.

Sen Gupta, B.K.S. 2003. *Modern Foraminifera*. Springer, The Netherlands, p 371.

Shi, X., Meng, X., Liu, G. et al. 2015. Annual variation of protozoan communities and its relationships to environmental conditions in a sub-trophic urban wetland ecosystem, southern China. Protistology, 9: 133–142.

Sibley, L.D., Hakansson, S. and Carruthers, V.B. 1998. Gliding motility: An efficient mechanism for cell penetration. Curr Biol, R8: 12–14.
Simarro, P.P., Cecchi, G., Franco, J.R. et al. 2012. Estimating and mapping the population at risk of sleeping sickness. PLoS Negl Trop Dis, 6: e1859.
Simonetti, A.B. 1996. The biology of malarial parasite in the mosquito—A review. Mem Inst Oswaldo Cruz, Rio de Janeiro, 91: 519–541.
Smith, G.M. 1944. A comparative study of the species of *Volvox*. Trans Am Microsc Soc, 63: 265–310.
Sokoloff, B. 1924. Das regenerations-problem bei protozoen. Arch Protistenk, 47: 143–252.
Sonneborn, T.M. 1939. *Paramecium aurelia*: Mating types and groups; lethal interactions; determination and inheritance. Am Nat, 73: 390–413.
Sonneborn, T.M. 1942. Sex hormones in unicellular organisms. Cold Spring Harbor Symposia on Quant Biol, 10: 111–124.
Sonneborn, T.M. 1947. Experimental control of the concentration of cytoplasmic genetic factors in *Paramecium*. Cold Spring Harbor Symp Quant Biol, 11: 236–255.
Stat, M., Morris, E. and Gates, R.D. 2008. Functional diversity in coral–dinoflagellate symbiosis. Proc Natl Acad Sci USA, 105: 9256–9261.
Stelzer, C-P. 2015. Does the avoidance of sexual costs increase fitness in asexual invaders? Proc Natl Acad Sci, USA, 112: 8851–8858.
Stephen, F.NG. 1990. Embryological perspective of sexual somatic development in ciliated protozoa: implications in immortality, sexual reproduction and inheritance of acquired characters. Phil Trans R Soc Lond, 329B: 287–305.
Stevens, D.A. 1998. Coccodiosis. In: *Encyclopedia of Immunology* (eds). Delves, P.J. and Roitt, I.M., Academic Press, London, pp 591–593.
Stoecker, D.E. 1999. Mixotrophy among dinoflagellates. J Euk Microbiol 24: 397–401.
Strauch, S., Becker, I., Polloth, L. et al. 2017. Restart capability of resting-states of *Euglena gracilis* after 9 months of dormancy: Preparation for autonomous space flight experiments. Int J Astrobiol, 17: 101–111.
Suzuki, N. and Not, F. 2015. Biology and Ecology of Radiolaria. In: *Marine Protists* (eds). Ohtsuka, S. et al. Springer, Japan, pp 179–222.
Takagi, H., Kimoto, K., Fujiki, T. et al. 2019. Characterizing photosymbiosis in modern planktonic Foraminifera. Biogeosciences, 16: 3377–3396.
Talman, A.M., Domarle, O., McKenzie, F.E. et al. 2004. Gametocytogenesis: The puberty of *Plasmodium falciparum*. Malaria J, 3: 24, http://www.malariajournal.com/content/3/1/24.
Taylor, F.R., Hoppenrath, M. and Saidarriaga, J.F. 2008. Dinoflagellate diversity and distribution. Biodivers Conserv, 17: 407–418.
Thurman, J., Parry, J.D., Hill, P.J. and Laybourn-Parry, J. 2010. The filter-feeding ciliates *Colpidium striatum* and *Tetrahymena pyriformis* display selective feeding behaviors in the presence of mixed, equally-sized, bacterial prey. Protist, 161: 577–588.
Tremaine, S.C. and Mills, A.L. 1991. Impact of water column acidification on protozoan bacterivory at the lake sediment-water interface. Appl Env Microbiol, 57: 775–784.
Umen, J. and Coelho, S. 2019. Algal sex determination and the evolution of anisogamy. Annu Rev Microbiol, 73: 12.1–12.25.
Umen, J.G. 2020. *Volvox* and volvocine green algae. EvoDevo, 11: 13, https://doi.org/10.1186/s13227-020-00158-7.
Undeen, A.H., Meer, R.K.V., Smittle, B.J. and Avery, S.W. 1984. The effect of gamma-irradiation on *Nosema algerae* (Microspora: Nosematidae) spore viability, germination, and carbohydrates. J Protozool, 31: 479–482.
Ushakov, B.P. 1968. Cellular resistance adaptation to temperature and thermostability of somatic cells with special reference to marine animals. Mar Biol, 1: 153–160.
Vd'acny, P., Ersekova, E., Soltya, K. et al. 2018. Co-existence of multiple bacterivorous clevelandellid ciliate species in hindgut of wood-feeding cockroaches in light of their prokaryotic consortium. Sci Rep, 8: 17749.
Verni, F. and Rosati, G. 2011. Resting cysts: A survival strategy in Protozoa Ciliophora. Ital J Zool, 78: 134–145.

Wang, W., Shor, L.M., LeBoeuf, E.J. et al. 2005. Mobility of Protozoa through narrow channels. Appl Env Microbiol, 71: 4628–4637.
Wang, Y., Seppanen-Laakso, T., Rischer, H. and Wiebe, M.G. 2018. *Euglena gracilis* growth and cell composition under different temperature, light and trophic conditions. PLoS One, 13: 3015329.
Warner, M.E., Fitt, W.K. and Schmidt, G.W. 1996. The effects of elevated temperature on the photosynthetic efficiency of zooxanthellae *in hospite* from four different species of reef coral: a novel approach. Plant Cell Env, 19: 291–299.
Warren, A., Esteban, G.F. and Finlay, B.J. 2016. Protozoa. In: *Thorp and Covich's Freshwater Invertebrates: Keys to Neartic Fauna*. Volume 2. (eds). Thorp, J.H. and Rogers, D.C., Academic Press, London, pp 5–38.
Wenzel, J., Gifford, C. and Hawkes, J. Economic impacts of trichomoniasis. https://beefrepro.org/wp-content/uploads/2020/09/Wenzel-Economic-Impacts-of-Trich.pdf.
WHO. 2015. *World Health Statistics*. World Health Organization, Geneva, p 164.
Wiser, M.F. 2010. *Protozoa and Human Disease*. Garland Science. Taylor and Francis, USA, p 256.
Wood, H.L., Spicer, J.I. and Widdicombe, S. 2008. Ocean acidification may increase calcification rates, but at a cost. Proc R Soc, 275B: 1767–1773.
Woodhall, D., Jones, J.L., Cantey, P. and Wilkins, P.P. 2014. Neglected parasitic infections: What every family physician needs to know. Am Fam Physician, 89: 803–811.
Woodruff, L.L. 1911. *Paramecium aurelia* and *P. caudatum*. J Morph, 22: 1–223.
Woodruff, L.L. 1913. The effect of excretion products of infusoria on the same and on different species with special reference to the protozoan sequence in infusions. J Exp Zool, 14: 575–582.
Woodruff, L.L. 1926. Eleven thousand generations of *Paramecium*. Q Rev Biol, 1: 436–438.
Yamamoto, K., Hamaji, T., Kawai-Toyooka, H. et al. 2021. Three genomes in the algal genus *Volvox* reveal the fate of a haploid sex-determining region after a transition to homothallism. Proc Natl Acad Sci USA, 118: e2100712118.
Yang, J. and Shen, Y. 2005. Morphology, biometry and distribution of *Difflugia biwae* Kawamura, 1918 (Protozoa: Rhizopoda). Acta Protozool, 44: 103–111.
Zakrys, B., Milanowski, R. and Karnkowska, A. 2017. Evolutionary origin of *Euglena*. Adv Exp Med Biol, 979: 3–17.

Author Index

G

H

I

J

K

L

M

Species Index

D

E

J

K

L

M

N

O

S

T

U

V

W

X

Y

Z

Subject Index

Author's Biography

Recipient of the S.S. Bhatnagar Prize, the highest Indian award for scientists, one of the ten National Professorships, T.J. Pandian has served as editor/member of editorial boards of many international journals. His books on Animal Energetics (Academic Press) identify him as a prolific but precise writer. His five volumes on Sexuality, Sex Determination and Differentiation in Fishes, published by CRC Press, are ranked with five stars. He has authored a multi-volume series on Reproduction and Development of Aquatic Invertebrates, of which the volumes on Crustacea, Mollusca, Echinodermata and Prochordata, Annelida, Platyhelminthes and Minor Phyla have been published. The CRC Press has recently published his new book series on Evolution and Speciation in Animals and Evolution and Speciation in Plants. The third volume on Evolution and Speciation in Protozoa is in your hands.

Author's Biography

Recipient of the S.S. Bhatnagar Prize, the highest Indian award for scientists, one of the ten National Professorships, T.J. Pandian has served as editor/member of editorial boards of many international journals. His books on Animal Energetics (Academic Press) identify him as a prolific but precise writer. His five volumes on Sexuality, Sex Determination and Differentiation in Fishes, published by CRC Press, are ranked with five stars. He has authored a multi-volume series on Reproduction and Development of Aquatic Invertebrates, of which the volumes on Crustacea, Mollusca, Echinodermata and Prochordata, Annelida, Platyhelminthes and Minor Phyla have been published. The CRC Press has recently published his new book series on Evolution and Speciation in Animals and Evolution and Speciation in Plants. The third volume on Evolution and Speciation in Protozoa is in your hands.